The Engines
of Our Ingenuity

An Engineer Looks at Technology and Culture

John H. Lienhard

OXFORD
UNIVERSITY PRESS

2000

OXFORD
UNIVERSITY PRESS

Oxford   New York
Athens   Auckland   Bangkok   Bogotá   Buenos Aires   Calcutta
Cape Town   Chennai   Dar es Salaam   Delhi   Florence   Hong Kong   Istanbul
Karachi   Kuala Lumpur   Madrid   Melbourne   Mexico City   Mumbai
Nairobi   Paris   São Paulo   Singapore   Taipei   Tokyo   Toronto   Warsaw

and associated companies in
Berlin   Ibadan

Copyright © 2000 by John H. Lienhard

Published by Oxford University Press, Inc.
198 Madison Avenue, New York, New York 10016

Oxford is a registered trademark of Oxford University Press

Library of Congress Cataloging-in-Publication Data
Lienhard, John H., 1930-
The Engines of Our Ingenuity: an engineer looks at technology and culture
/ by John H. Lienhard.
p. cm.
Includes bibliographical references.
ISBN 0-19-513583-0
1. Technology—Social aspects.   2. Creative ability in technology.   I. Title.
T14.5.L52 2000
303.48'3—dc21   99-37614

Four lines of "The Man with the Blue Guitar" are from *Collected Works* by
Wallace Stevens. Copyright 1936 by Wallace Stevens and renewed 1964 by
Holly Stevens. Reprinted by permission of Alfred A. Knopf, a Division of
Random House, Inc.

Book design by Adam B. Bohannon

9 8 7 6 5 4
Printed in the United States of America
on acid-free paper

Contents

Preface

It would seem only reasonable that, to tell the story of technology, one must first learn something about that great sprawling enterprise. In my case, I worked for a half-century as an engineer and that was after spending a childhood building model airplanes and homemade replicas of all the dazzling new twentieth-century machines. Then I read a whole library-full of books about technology.

All that has left me with a keen understanding of the blind beggars who came upon an elephant. As those fabled beggars stumbled into various parts of the elephant—its side, its ear, its trunk—one beggar thought the animal was a wall, another a leaf, another a serpent. We too might as well be blind when we meet the elephant of technology, for it is far too large for any of us to see whole.

What you will find here is a progression of seventeen glimpses of the creature, seventeen chapters describing what I have known of the elephant's foot, its tail, its tusk. Of course there is so much more of the elephant than anyone could ever hope to contain in a book. Every overview of technology ever compiled has necessarily been arbitrary—fine accounts of the elephant's eye or of its leg.

So I make no apologies for the particular roll of the dice that made this book. It is based loosely on the first year's broadcasts of my daily Public Radio series, *The Engines of Our Ingenuity*. Those programs have dealt with human ingenuity and creativity seen largely through the window of history, and they set the tone of the series ever since. I have organized those early scripts into chapters, throwing some programs away and adding a few from later in the series to flesh out ideas. Then I rewrote each set to form it into a single essay on one facet of our technological enterprise.

The result is not meant to be a work of history. Insofar as I deal with history, the book is largely derivative. It is, instead, commentary and autobiography. (Though autobiography always lurks in the shadows, I keep it muted until the last chapter.) I try to convey the texture of the elephant's tail in one chapter and of its tusk in another. And I try to tell what it means, at last, to bond with the beast.

I stress in Chapter Fourteen that no good work—no invention and no book—is one person's doing. This book is filled with the good will of hundreds of people whose contributions flow through it: KUHF-FM radio station personnel, colleagues, librarians, university administrators who have supported the work, listeners from around the world (Armed Forces Radio and the BBC have carried *Engines* programs internationally), friends and family. I identify contributors to the radio program in *The Engines of Our Ingenuity* web site where complete transcripts (and more) may be found. While these people have also contributed in spirit to the book, I restrict the credits below to those people who have helped bring this book into being.

The idea of making the scripts into a book was first suggested by Bill Begell, then president of Hemisphere Publishing Corp. The project was begun in 1989, but it eventually perished when Hemisphere passed to companies whose interest was in technical textbooks and handbooks. Peter Gordon, then president of Textware Corp., took the project over in 1996. His energy, enthusiasm, and critical edge breathed new life into it. When Peter closed his company and took a senior editorship with Oxford University Press, he took the project with him.

It was on the advice of Kirk Jensen at Oxford University Press that I changed the format from a set of stand-alone, 500-word scripts to a loosely connected set of seventeen essays. Such a format better suits a reading audience just as the short scripts suit listeners. I am grateful to him and to the many people at Oxford who have finally made this work into a book. Susan Day managed the editorial task and provided countless improvements in the text.

I am also deeply indebted to many friends who have critically read chapters of the manuscript. First among these are two people: Carol Lienhard has not only turned her discerning eye upon every word in the book, but on all the radio scripts that preceded it; Dr. Jeff Fadell, linguist and librarian at the University of Houston, has edited both the manuscript and the entire set of *Engines* scripts. These two very different views have played formative roles in shaping what you will find when you turn this page.

John H. Lienhard
September, 1999

1

Mirrored by
Our Machines

A mirror is a strange device. Stand in front of one and what comes back is not the *you* that you know. Rather, it is you, turned about and shown to you in a crazy literal way. You see the exact reverse of what others see when they look at you. If a photographer hands you a picture (taken in just the right way) of me standing before a mirror, you might have a hard time telling which is the reflection and which is the reality. Mirrors put us off balance by being both literal and subtle at the same time.

When I call our technologies mirrors of ourselves I do not do so lightly. An alien looking at Earth for the first time would certainly seek to know us by gazing upon our reflection in our machines. Indeed, that is what anthropologists do when they examine the alien skeletons of our ancient forebears. Before anthropologists identify a particular primate skull as human, they search the area where they found it for evidence of toolmaking.

The very word *technology* helps us understand this process. The Greek word τεχνη (or *techne*) describes art and skill in making things. Τεχνη is the work of a sculptor, a stonemason, a composer, or an engineer. The suffix *-ology* means the study or the lore of something. Technology is the *knowledge* of making things. Some people have argued that we should not call our species *Homo sapiens*, "the wise ones," but rather *Homo technologicus*, "they who use τεχνη," for that is who we are.

There is more to τεχνη+ology than that. We freed our hands by walking on our hind legs before we took up toolmaking, and we made our earliest stone tools some 2.4 million years ago, while our skulls still accommodated a relatively small brain. Our capacity for thought began to grow as we created increasingly sophisticated implements. Technology has driven our brains. Our expanded physical capabilities made technology—extended toolmaking—inevitable. Technology has, in turn, expanded our minds and fed itself.

At first, the notion that technology drives our minds may be surprising. Shouldn't it be the other way around? After all, we teach people to be technologists. We train their minds so they can dictate the course of technology. Yet who on this planet would be clever enough to invent, say, a microcomputer? Who *did* invent the microcomputer? The answer is that *nobody did.* It invented itself! At each point in its evolution the machine revealed more of its potential. In each stage it exposed one more step that this or that person recognized and leaped to complete.

One Christmas my wife and I gave a primitive Vic20 home computer to our then-fifteen-year-old son. He vanished into his room for two weeks and emerged about Epiphanytide (appropriately enough), able to program in Basic. Who taught him? The computer did. It expanded his mind and made him more than he was. He came out of his room changed. Like all of us, he was being shaped by his technology. Technology, the lore of making and using implements, is a primary element in our cultural heritage. The tools, implements, and machines around us enfold and instruct us from birth to death.

In this sense I am hardly guilty of hyperbole when I say that the computer invented itself. We instinctively build machines that resonate with us. The technologies of writing and printing each altered the way we see the world. Each opened our eyes to the expanded possibilities they presented to us. Each profoundly changed our civilization.

The automobile led to things that never crossed the minds of its inventors. It led us, for good or for ill, to invent highway systems and to change the form of cities. The invention of the telephone altered the texture of human interaction. When someone asked Wilbur Wright the purpose of his new flying machine, he answered, "Sport, first of all." The airplane had yet to serve as a mirror that would reveal our deeper dreams and needs.

So we begin to understand technology and humankind when we step through the mirror of our machines, when we weigh the chicken-oregg question of whether the mind drives technology or technology drives the mind, and when we look at the way the existential fun of making things is born in the interaction between our own inventiveness and the technology that surrounds and drives our thinking.

A good place to start is with one of the most important of all human technologies: farming. Farming was one of the great steps we had to make on the way to fulfilling our destiny as a species of builders and makers. We had to leave hunting and gathering and assume control of the wealth of the land.

The key circumstance triggering that great leap forward was a remarkable pair of genetic accidents. Archaeological evidence shows us the two events that set the stage. Before 8000 B.C., the ancestor of wheat closely resembled a wild grass rather than the rich grain-bearing plant whose seeds we eat today. Then a genetic change occurred in which this plant was crossed with another grass. The result was a fertile hybrid called *emmer*, with edible seeds that blew in the wind and sowed themselves.

In 8000 B.C. a hunting-gathering people called the Natufians lived in the region around Jericho and the Dead Sea. By then, the climate had been warming for two thousand years. Once the area had been fairly lush. Now it grew arid. Game moved north and the vegetation changed. The wild grains grew well in a drier climate, and the Natufians began shifting their diet toward grain. They took to harvesting and eating the emmer seeds. But they didn't have to worry about planting emmer because it sowed itself.

The second genetic accident occurred sometime before 6000 B.C. It yielded something very close to our modern wheat, with its much plumper grain. But the new grain, even if it was fertile, could not survive on its own. Although wheat is far better food than emmer, its seeds don't go anywhere. They are bound more firmly to the stalk and are unable to ride the wind. Without farmers to collect and sow wheat, it dies. Modern wheat created farming by wedding its survival to that of the farmer, and it left us with a great riddle: How did modern wheat replace those wild grains? We don't know, but we can guess.

The Natufians probably reached the point of planting their own emmer to supply the grain they needed. Once they did, the fat wheat had its chance because it was easier to harvest. Its seeds don't blow away when you cut it. Every time the Natufians harvested seed, they got proportionately more of the mutations and lost more of the wild grain.

Top: Modern white Gaines wheat.
Middle: Emmer.
Bottom: A wild wheatlike grass, *triticum monococcum*.

It took only a generation or so of planting before the new grain took over. In no time at all, modern wheat dominated the fields. That was both a blessing and a curse. The Natufians unwittingly replaced the old wild wheat with a higher-yield crop. But it was a crop that could survive only by their continued intervention. No more lilies of the field! From now on we would live better, but we would also be forever bound to this new food by the new technology of agriculture.[1]

And so the technology of farming mirrors the farmer. Humans created farming, and farming made humans into something far different from what they had been. The process was no different than my son's interaction with that primitive computer.

We can see just how deeply this process of mirroring runs through all our technologies (and the sciences that, as we shall see, have been built upon those technologies) if we look at units of measurement. When Protagoras said that "Man is the measure of all things," almost twenty-five hundred years ago, he was closer to literal truth than we might at first think. The gauges and meters we use to measure things usually begin by copying our own senses. All our weights and measures, in some way or another, reflect what we see and feel.

A pound or a kilogram, for example, is roughly the mass of any fairly dense material, like a rock or a piece of metal, that we can hold comfortably in our hand. The inch, foot, yard, and meter all correspond roughly with various body parts. The mile and kilometer also have a meaning that is made clear in parts of rural America where people talk about the distance of a "see." Ask someone in eastern Kentucky how far it is into town and he might say, "Oh, 'bout two sees." He means you should look down the road as far as you can see. Where your vision runs out, you spot, say, an oak tree. You walk to it and look again. There, in the distance, is the town, just two sees away. How far is a see? Of course it varies. But even in flat terrain our ability to make things out usually ends after about a kilometer or, at best, a mile.

We divide thermometers into degrees Fahrenheit or Celsius, and these are roughly the smallest increments of temperature we can feel. We usually know if we have a one-degree fever. We can sense about one volt with our tongue; our ears are sensitive to about one pound per square inch of pressure change; and so on.

Mirrored by Our Machines

The units of a kilowatt or a horsepower represent roughly the power most of us can produce in a short sprint—like running upstairs. By the way, the unit of a horsepower is less than the short-term work of a real horse, but considerably more than a horse can sustain all day long. Not only is the kilowatt or horsepower close to the maximum power you or I can produce in a short burst; it is also the most power we can tangle with without being hurt. They represent about as much energy as the sun pours on us if we lie on the beach at midday, or the rate at which we consume energy when we take a hot shower.[2]

Since we are the basis for most measuring devices, science reflects the world in human terms. But that is not really so bad. Most scientists know perfectly well that science has not yet reached ultimate truth of any sort. The work of science nevertheless yields constructs that make our experiences predictable. Today's science-based engineering obviously has to mirror human needs and human nature. And so does science itself.

Still, the immediate reflection of our own bodies in the physical measures that we use every day leaves us struggling at each turn to see more objectively—to shake off human limitations. The magnitude of that problem emerges when we pose a deceptively simple question, "Should we regard a certain object as big or small?" To answer, we instinctively refer to the size of our own body. We understand size on the scale we experience it, and we can be surprised by how differently an object will behave when it is much larger or smaller than our bodies.

To see what I mean, you might try this experiment: First find a very large metal sphere and a very small one—say, the big steel ball used in the shot put and a BB. Now drop each from a height of a few feet into a swimming pool. You will see that the shot splash is not at all like a scaled-up BB splash. The large shot sends out a sheet of water that breaks into a fine spray of drops. There are only a few drops in the BB splash. In fact, that's how we know whether the naval battle in a movie is a scale model or footage from a real battle. The splashes look wrong in the scale model.

I once knew a badly crippled construction worker. He had been working on the ledge of a building that was being demolished when he saw a two-ton scoop swinging toward him—very gently, very slowly. He put out his hands to stop it as he might have stopped a child on a swing, and when it reached him, it very gently crushed him. His experience with playground swings had grievously misled him about the behavior of two-ton scoops.

Engineers think a lot about making scale models of big prototypes. We would not get very far if we had to make full-size wind tunnel tests of a Boeing 747. The trick is to set the conditions in a small model so that its behavior is similar to the large prototype. We really *could* use a BB experiment to learn what a large shot does when it hits water if we changed two things. The BB would have to move much faster than the shot, and we would have to put just the right amount of detergent in the water to cut its surface tension.

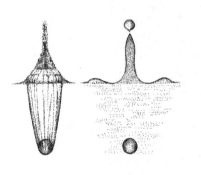

The forces that dominate this process (inertia, gravity, and surface tension) all vary in different ways with size. The theory of modeling tells how to stretch the dimensions of the relevant variables into universal values. When we do that, surprising things happen. Suppose, for example, that we want to scale up instead of scaling down. Suppose we want to study the movement of microorganisms in our body fluids using laboratory experiments in the visible world. Modeling theory tells us we can do this if we stretch time and magnify liquid resistance. We can replicate the motions of blood cells or spermatozoa by moving large models of these organisms very slowly through cold honey.

Sir George Cayley, born in 1773 in Yorkshire, came to a remarkable insight about scaling physical phenomena when he was a young man trying to solve the age-old riddle of human flight. Cayley made a number of key discoveries, but none was more surprising than his realization that trout have the ideal, minimum-resistance, body shape for an airplane.[3] Why a trout and not a bird? It is because the flow of water around a fish and the flow of air around a bird of the same size are very different. A century later we had the rules of dynamic similitude. They show that when we scale the interactions of viscous and inertial forces, a small fish in water moves far more like a large machine in the air, than a small bird in the air does. That's why the design of subsonic airplanes eventually settled on a shape far more like fish than birds. Our machines still mirror our experience, but now that experience is tempered by scientific theory.

So the problem of modeling is one part of a general problem we face whenever we design things. We have to find ways to see what is not obvious to our eyes. We have to find ways to predict complicated behavior before it becomes part of our experience. Our modern systems of weights and measures evolved as scientific instruments gave increasing precision and definition to measurement. However, technology reached high levels of sophistication long before we had any such apparatus.

My grandmother used to tell me that if I burned my finger, I should dip it in a cup of tea. She knew that before doctors knew anything about the healing power of the tannic acid in tea. My grandmother's finely honed intelligence was in no way lessened by the fact that she'd never studied organic chemistry.

Take the ancient technology of Japanese sword making, which reached an astonishing perfection twelve hundred years ago. A samurai sword is a wonderfully delicate and complex piece of engineering. The steel of the blade is heated, folded, and beaten, over and over, until the blade is formed by 32,768 layers, forge-welded to one another. Each layer is 0.00001 inch thick. All that work was done to very accurate standards of heat treatment. The result was an obsidian-hard blade with willowlike flexibility.

The blades represented a perfection of production standards that has yet to be matched by modern quality control. The Japanese craftsmen who made them knew nothing about temperature measurement or the carbon content of steel. How do you suppose they got it right, again and again?

The answer is one we are well advised to remember. Sword making was swathed in ceremony and ritual. It was consistent because the ceremony was precise and unvaried. Heat-treating temperatures were controlled by holding the blade to the color of the morning sun. The exact hue was transmitted from master to apprentice down through the centuries. Sword making was a part of Japanese art, and it was subsumed into Japanese culture.[4]

That form of quality control was not unique to the Japanese. It was true of eighteenth-century violin making and it is still true in other older technologies that survive today. Ritual can do much of what we do with weights and measures. Our intelligence, after all, runs deeper than our ability to read gauges. Great technologies arise out of a full range of experience. They come from creativity triggered by more than tables of technical data. Good technology is not independent of culture. The best doctor knows organic chemistry as well as his grandmother's folklore. The best metallurgist knows about iron-carbon phase diagrams but can see those diagrams in the light of a bright yellow-orange blade emerging from the forge.

Years ago I worked for a seasoned design engineer. One day he looked at a piece of equipment and said, "Look at that heavy gear running on that skinny shaft. Some designer didn't use his eyes." The best engineers know math, physics, and thermodynamics, but they also know the world they live in. The best engineers bring a visceral and human dimension to the exacting math-driven, science-based business of shaping the world around us. The machines they produce therefore mirror themselves.

The easiest place to see the mirror of technology is in the language we use to talk about our technologies. The words *science, technology*, and *engineering* take a terrible beating. Who makes a spaceship fly—a scientist, a technologist, or an engineer? Who should shoulder the blame if it fails? These questions are easier to answer if we really understand what the words mean.

The word *science* comes from the Latin word *scientia*, which means "knowledge." We apply the word *science* to ordered or systematic knowledge. A scientist identifies what is known about things and puts that knowledge into some kind of order.

We have noted that the word *technology* combines the Greek word τεχνη (combined art and skill) with the ending *-ology* (the lore or the science of something.) In its role as the science of making things, technology stands apart from the actual act of glassblowing or machining. It is the ordered knowledge of these things. It is also our instinct for sharing our knowledge of technique. Our language would be a lot clearer if we could reclaim the old Greek word τεχνη and restrict its use to describing the actual act of making things.

The last of the three words, *engineering*, comes from the Latin word *ingenium*. That means "mental power." English is full of words related to *ingenium: ingenuity*, which means "inventiveness," and *engine*, which can refer to any machine of our devising—any engine of our ingenuity. So an engineer, first and foremost, devises machines. For about three hundred years science and τεχνη have joined forces. We talk more about that in chapter 5. Today's engineers are technologists who are well schooled in science and can make effective use of it when they try to create the engines of their ingenuity.

Which of the three, scientist, technologist, or engineer, deserves the credit for the success, or blame for the failure, of a spaceship? The answer, of course, is that the question is no good. The three functions of τεχνη, science, and invention work together to make a spaceship. Engineers combine these functions. One engineer might behave more like a craftsman—a user of τεχνη — while another might behave more like a scientist, refining background information for designers. But people earn the title *engineer* when the *goal* of their labors is the actual creative design process—when they combine a knowledge of τεχνη with science to achieve invention.

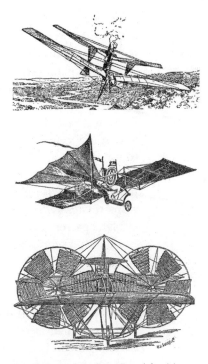

Top: Langley's Aerodrome. Center: Henson's Aerostat. Bottom: Moy's Aerial Steamer.

Look further at words, at the way we name our machines. A machine normally receives its permanent name only after it has achieved a certain level of maturity— after it has settled itself into our lives. Take the airplane. A hundred years ago, we had dozens of terms to describe it: *aerial velocipede, aerial screw machine, aero-motive engine, bird machine,* and *flying machine*. Most of these names vanished ten years after the Wright brothers flew. Now we have settled on just two terms, *airplane* and *aircraft*.

No one I knew had a refrigerator when I was little. We had an *icebox* with a rack on top where we put a new fifty-pound block of ice every few days. I still forget, and annoy my sons, by calling our refrigerator an icebox. During the 1930s we tried all kinds of terms for the new machine—*Frigidaire, electric icebox,* and of course *refrigerator*.

The words *engine* and *machine* show up repeatedly when devices are first named. They are from Latin and Greek roots and broadly refer to devices that carry out functions. The steam engine was first called a *fire engine,* and it still keeps the *engine* part of its name. We still say *sewing machine,* but no one calls a telescope an *optical engine* anymore (as they did in the seventeenth century). I especially like the name Babbage gave his first programmable computer in the early eighteenth century. He called it an *analytical engine*. Software packages for checking programs were called *parsing engines* long before another engine word attached itself to computers: the now-common term *search engine*.

Foreign names stick to new gadgets for a while, but they tend to fade. Airplane designers have moved away from the French words *empennage, fuselage*, and *nacelle* in favor of the English equivalents: *tail, body*, and *pod*. The German name *zeppelin* was given to one form of what the French call a *dirigible*. Nowadays we are increasingly inclined to use the English word *airship*. We call a writing desk an *escritoire* only when we want to run up its price. The first names we give new technologies often tie them to older ones. An early name for the first airships was *aerial locomotive*; railway passengers still ride in *coaches*; and airplane passengers pay *coach fares*.

Finally, a game we all might play: Over the next decade, track the changing computer-related names. Watch as we run through words such as *screen, CRT*, and *monitor*, or *Internet* and *Web*. Watch us select among names like *microcomputer, PC, workstation*, or simply *the machine*. Watch as those systems become metaphors for who we are. When we finally settle on names, what we shall really be doing is taking the machine fully into our lives.

Thus far, I have described the mirror of our technology in fairly objective terms, but technology lies too close to the human heart to be dealt with in such a straightforward way. The mirror reflects aspects of our nature that are not immediately obvious to ourselves. I can clarify my meaning here by asking yet another question: What is the oldest human technology?

Farming developed late in human history. Before farming, settled herdsmen and gatherers made clothing, knives, tents, and spears, but so did nomads before them. Go back further: Archaeologists show us that pictures and music were among the Stone Age technologies. Magnificent cave paintings have survived since the beginning of the Upper Paleolithic period—at least twenty-five thousand years ago. Along with them we found evidence of rattles, drums, pipes, and shell trumpets. Even the Bible, the chronology of the Hebrew tribes, identifies the making of musical instruments as one of three technologies that arose in the seventh and eighth generations after Adam.

Music is clearly as old as any technology we can date. Couple that with the sure knowledge that whales were composing complex songs long before we walked this earth—that the animal urge to make music precedes technology—and I offer music making as my candidate for the oldest technology.

The societies with the least technology still make sophisticated music. Song, dance, musical instruments, and poetry are central to ancient Australian Aborigine culture. Music is the most accessible art and, at the same time, the most elusive. In almost any age or any society, music making is every bit as complex as other technologies. But our own experience tells us as much as archaeology does. Experience tells us that music is primal. It is not just a simple pleasure. Jessica says to Lorenzo in Shakespeare's *Merchant of Venice*: "I am never merry when I hear sweet music." Lorenzo replies,

> The reason is your spirits are attentive, ...
> The man that hath no music in himself, ...
> Is fit for treasons, ...

And we know what he means! If we cannot respond to art, to music, then something *is* missing and we *are* fit for treasons. Music helps us understand the human lot. Music is as functional as any worthwhile technology. Its function is to put reality in terms that make sense. That means dramatizing what we see—transmuting it into something more than is obvious. Poet Wallace Stevens wrote:

> They said, "You have a blue guitar,
> You do not play things as they are."
> The man replied "Things as they are
> Are changed upon the blue guitar."[5]

The blue guitar—music, or any art—does change reality. It turns the human dilemma around until we see it in perspective. Sometimes it takes us through grief and pain to do that, disturbing us at the same time it comforts us. But it serves fundamental human need. So it is no coincidence that the technologies for creating art, and music in particular, preceded all else.

Those subjective factors are always at work, shaping technologies to serve us best—shaping them to serve more elemental needs than are evident. The reason technology is impossible to predict is that our predictions are inevitably shaped by those factors that are fairly evident. Imagine for a moment that the year is 1800 and we are called upon to predict the nature of American culture once it has moved west of the Mississippi. I doubt any of us could have come close.

What actually happened was that the American West developed highly characteristic technologies for daily life. We all know the texture today: log cabins, windmills, card games, heavy horse-drawn wagons, whiskey, large saddles, and (I might ominously add) death by hanging.

Historian Lynn White pointed out a startling feature of all these technologies: their link to the Middle Ages. Log cabins were a medieval form of housing; the earlier Romans and later Europeans used much different building technologies. The Romans and later Europeans drank beer and wine, but whiskey was the medieval drink of choice. Romans and eighteenth-century gamesmen used dice, but you would find only cards in medieval or western saloons. That sort of comparison can be made right down the line. The Romans executed people by crucifixion and the later Europeans used beheading and shooting, but strangulation—hanging—was the standard medieval punishment.

The strange parallel grows more puzzling when we learn that the middle-class settlers of New England tried to re-create what they had left behind, instead of looking for the most efficient technologies. They tried to go straightaway to the beam-and-plank method of house construction they had used in England, even when log houses made better sense.

But the settlers of the West were generally from the European lower classes. They were peasants, workmen, and people who had lived away from the sophisticated centers of Europe. Their lives had generally been closer to the technologies of the Middle Ages. These people found their way more quickly to the sort of rough-hewn ways that worked so well in both the medieval world and the undeveloped West. They also held little nostalgia for the current styles in Europe.[6]

White's suggestion bothered many other historians. He did not explain the similarities, and people do not like what cannot be explained. Yet, in a way, it seems fairly evident. The technologies of the eleventh through the fifteenth centuries were wonderfully direct, practical, and inventive, and so too were the immigrants to the American West. Medieval life and western life were open to variety and change. What the Old West really did was to mirror medieval life accurately, because it was populated by free and inventive people who knew how to adapt to new circumstances. They allowed light to pass through that mirror, and the result was one that was totally unpredictable.

Once those technologies were in place they had a staying power that allowed them to endure long after we might have expected them to have fallen by the wayside. The western farm windmill is an icon that will not easily yield to small electrical pumps. The playing cards remain. Cowboy clothing has been stylized and stays with us long after we've forgotten its distinct purposes.

Case in point: the old railway train caboose. The caboose was a kind of moving rear observation tower—a way of seeing over and beyond a railway train. Electronic safety systems have made them obsolete, but cabooses are so woven into the fabric of railroading that in 1980 railroad companies were caught up in a bitter debate as to whether they should be abandoned.

So goes the story of lighthouses as well. Lighthouses call up the romance of the sea just as powerfully as cabooses complete the image of railroading in our mind's eye. Lighthouses are used to mark all kinds of dangers to shipping at night—sea crossings, rocks, major landfalls. Height is important: a 15-foot-tall light can be seen four and a half miles away, a 120-foot-tall light is visible for more than twelve miles, and so forth.

That's why the Pharos at Alexandria was so large. One of the seven wonders of the ancient world, it held a huge bonfire four hundred feet in the air. And, like it, lighthouses through the centuries have usually been tall masonry towers mounted on the shore, or maybe on shoreline cliffs, burning wood, olive oil, or, later, coal or candles. It was a pretty static technology. Rotating beams were not invented until 1611. It was 1763 before reflectors were finally placed behind lamps to boost their power. The first lens was introduced less than two hundred years ago.

Lighthouse construction started moving again in 1698 when the English had to warn ships away from the Eddystone rocks, fourteen miles southwest of Plymouth. You might well know about the lighthouse that was eventually put there through this old folk song:

Oh, me father was the keeper of the Eddystone light,
And he slept with a mermaid one fine night.
And from this union there came three,
A porpoise and a porgy—and the other was me.

Building the Eddystone light was a terrible job that greatly advanced the whole technology. It had to be erected right at sea level, where it was hammered by waves. The first one, made of wood, lasted five years. The next, made of wood and iron, burned down after forty-seven years. The great English engineer John Smeaton designed and built the third Eddystone lighthouse in 1759. He used a new kind of interlocking stone construction that was not replaced by the present Eddystone light until 1881.

But now radar, sonar, and electronic buoys are putting an end to the lighthouse. We will have to live in a world without cabooses on trains, and without those beautiful storm-beaten minarets to call the weary sailor home. The siren attraction of the lighthouse, like other technology past (and, I suppose, like much technology yet to come), is that good technology is contrived to fulfill human need. That is why it satisfies more than function—why it expresses what is inside us. All good technology acquires symbolic as well as functional power. That is the reason we are so loath to say good-bye to it.

Sometimes the metaphor races ahead of function. That happens today, but it is usually too hard to sort out while it's happening. Go instead back to the world of the 1930s. The watchword in those days was *modern*. If I knew one thing as a child, it was that I lived in the modern world. It was a world where the vertical lines of art deco were giving way to horizontal streamlined forms. Everything in the 1930s, '40s, and '50s was streamlined. The Douglas DC-3 had brought streamlining to passenger airplanes. The Chrysler Airflow and the Lincoln Zephyr brought it to the automobile. Even my first bike was streamlined.

Earlier in the twentieth century, the great German experts in fluid flow had shown us how to streamline bodies to reduce wind resistance. Streamlining certainly served this function when things moved fast. But my bicycle hardly qualified. Nor did the streamlined Microchef kitchen stove that came out in 1930. Bathrooms were streamlined. Tractors were streamlined. Streamlining was a metaphor for the brave new world we all lived in.

A confusion of design schools competed with each other in the early 1930s. The German Bauhaus school had been scattered by the Nazis. Art deco was dying. Neither the classic colonials nor Le Corbusier and the other architects associated with the International Style could gain ascendancy.

Then streamlining came out of this gaggle, propelled by American industry and making its simple appeal to the child in all of us. It certainly appealed to the child I was then. Of course, streamlining was in part a sales gimmick—something to distract us from the tawdry realities of the Depression. It told us to buy things. It told us we could all go fast. It was hardly one of the great humanist schools of design.

The Nazis and Bolsheviks used streamlining as a propaganda tool. American industry used it to make us into consumers. It hinted at the daemon of technocracy. It lasted until the 1950s, when its dying excesses—enormous tail fins and chrome bumpers—made the automobile seem ridiculous by most esthetic standards.[7]

But I loved airplanes as a child, and the functional curved aeroform shape touched something in me. The way the gentle camber of an airfoil gave the invisible wind a handle by which to pluck a fifty-ton airplane into the sky—that was truly magical.

"When I was a child," Saint Paul said, "I thought as a child." Streamlining may well have been a childish symbol of our modern world, which we have now put away with other childish things. But I still look back with suppressed joy at that vision of motion, speed, and buoyancy—at the DC-3, the 1949 Studebaker, even big tail fins. That child is still in me.

Perhaps the mirror becomes most treacherous when machines converge with human perception. Professor Kenneth Torrance of Cornell University gave a talk back in 1988 that inadvertently dramatized this theme. He began by using computer graphics to create simple scenes—a crib in a room, a jar on a table. To illuminate his scenes, he let a computer chew through the complicated equations he had written for the reflection and diffusion of light until the image was lit the same way a lamp or the sun would light it.

By then we had seen the first computer-generated images in the movies, but his were much better. Colors appeared just the way they should. They mixed perfectly in the shadows. These pictures had the beauty and accuracy of a Dutch master's work. When Torrance had finished, I was no longer sure whether I was looking at a picture or at the thing itself. These were not just artist's creations. Torrance had written the rules of nature and then let the computer obey those rules. In a sense, he had told the computer how to re-create the actual *being* of nature.

Of course, it is not easy to parse reality into the language of computers. Yet when we do, the results are not just stunning, they are disorienting as well. Students of fluid flow struggle to make their computers tell them how fluids move over airfoils, through tubes, past turbine blades. As computers replicate the tortuous swirls of water and air in slow-motion, on a computer screen, we sometimes wonder whether we are seeing reality or the imaginings of a lunatic.

By now, the science of computer graphics has moved far ahead and we see stunning machine-generated realities on movie screens or even on our computer monitors. Today's scientist can do many experiments more accurately within the computer than in the laboratory. Yet some computer modeling is terribly deceptive. Its seeming accuracy can miss features that would have been revealed in the laboratory. In either case, as the computer's role expands, we users adopt the language of people dealing with real things. We speak of doing "numerical experiments" when we isolate processes on the machine instead of in the laboratory. We can be disarmingly casual about separating computer and laboratory data. The computer takes a larger and larger role as a partner in human decision making.

We no longer can be sure who created the picture we are looking at: an artist, a camera, or a computer. The computer can replicate the sound of a concert grand piano and fool me into thinking that I hear a live person playing a real instrument. As the computer speaks to our senses as well as to our intellects, we start to have trouble finding the line between realities of the machine and realities outside it.

Machines mirror our lives. Our lives mirror our machines. We've seen how devices change us. Machines extend our reach and take us where our legs cannot. They amplify our voices. They even give us wings.

I talk about machines extending our bodies because that is the way they touch us so powerfully. But replacing our legs with an automobile, or our backs with a forklift, is nothing compared to what computers do. They sit right beside our brains and assume a kind of partnership practically inside our heads. Our relations with machines have always been personal, but with our computers they are terrifyingly so. Just how personal has come home to me in a very real way in recent years.

Between 1983 and 1988 I wrote everything on my first word processor—papers, talks, letters, and two books, well over a million words of finished copy and several times that in discarded drafts. Imagine, if you can, my intimacy with that machine when, after five years, it became clear that it had grown obsolete. By then it held my thoughts and was giving them back to me. But time had now passed it by.

I had no choice but to buy a flashy new computer. Once I did, it worked diligently to further change me. It had ten times the memory of the old machine, two hundred times the storage capacity. It had a colorful new screen. It thought with blinding speed. Now it played chess and Othello with me. It handled several manuscripts at the same time, corrected my spelling, indexed my texts, and tended my files and addresses. It suggested better words for me to use. By the time I was done with the second computer and moved on to a third one with ten or a hundred times more capacity, the second one still held countless surprises I had yet to discover.

By now I had realized that each of those new computers knew all the tricks of behavior modification. If I said the wrong thing, the machine stopped talking to me and feigned ignorance. It confided its secrets to me only if I said just the right words to it. During the first month each new computer has kept me on the rack, rewarding me now and then by tossing me a new bone.

After a month, the transition begins to complete itself. The new machine is not yet the comfortable old shoe that the one before it had become. But it gets there, and its way of getting there is by changing me. It is in the transition from one machine to another that we come to appreciate their power in our lives. Do you remember your first bike or the first car you drove? Think back for a moment. That bike was like a flying carpet. It changed you, irrevocably.

People often ask, "Do these transitions occur for good or ill?" But that's not very helpful, because machine making is an inseparable part of us. We are mirrored by our machines, and the corollary is also inescapable: We mirror our machines. The question is not whether we should let them change us, but whether we are to be lifted up or dragged down in the process. That issue hovers over the rest of what follows as we talk about technology and its place in our lives.

2

God,
the Master Craftsman

Adam awoke on the eighth day of creation, measuring his newly gained creative powers. In a harsh, forbidding world, somewhere to the east of Eden, Adam flexed new muscles and smiled. "That garden was nothing," he chuckled. "We're well rid of it. I'll build a garden that'll put it to shame."

That eighth day of creation was, in fact, very late in time. Adam had hunted and gathered in the garden for four million years. Then, just the other day—only about thirty thousand years ago—he came into the dense, self-reinforcing, technical knowledge that has, ever since, driven him further and further from the garden.

We are a willful, apple-driven, and mind-obsessed people. That side of our nature is not one that we can dodge for very long. Perhaps the greatest accomplishment of the eleventh-century Christian church was that it forged a tentative peace with human restlessness. All the great monotheistic religions of the world have honored God as Maker of the world, but the medieval Christian church went much further: It asserted that God had manifested himself in human form as a carpenter—a technologist, a creator scaled to human proportions. It seemed clear that if we are cast in God's image, then God must rightly be honored as the Master Craftsman.

The peace forged between the medieval Church and Adam's apple was wonderfully expressed by an anonymous fourteenth-century Anglo-Saxon monk who sang:

Adam lay ibounden, bounden in a bond.
Four thousand winter, thought he not to long.

And all was for an appil, and appil that he tok,
As clerkès finden, written in their book.

Ne had the appil takè ben,
Ne haddè never our lady, a ben hevenè quene.
Blessèd be the time that appil takè was
Therefore we moun singen,
Deo Gracias!

In the mind of that monk, taking the apple of technological knowledge was the first step in spinning out the whole tapestry of the biblical drama that had left him at last with the comfort of his Virgin Mary. So he sang "Deo Gracias," picked up his compass and square, and went to work. No wonder medieval Christianity was such a friend to the work of making things.

The great emergence of Western technology can be traced from the medieval Church straight through to the Industrial Revolution. That central fact is easy to forget in light of today's technology, interwoven as it is with a science dedicated to detachment and objectivity. The ideal of objective detachment has evolved only over the last three hundred years, and it misleads us. It causes us to see technology as unrelated to driving forces deep within the human heart.

But such detachment is not only very young; it is also something modern engineering designers have begun to question. Designers today are consciously trying to tap into subjective thinking in the creative process. They are increasingly aware that invention is driven by imperatives from deep within us. Historians have begun to see that the technological empire grew out of spiritual seeds sown in medieval Europe.

Late-twentieth-century thinking portrays objective science and technology as being, at best, neutral on religious matters. At worst, technology and religion have been portrayed as adversaries. That antagonism grew up as the new experimental sciences began to conflict with literal readings of the Bible. But the quintessential twentieth-century scientist, Albert Einstein, simply gazed at a God whom he acknowledged in only the most abstract terms, and said of his God, "Subtle is the Lord."

Modern physics has proven to be subtler than modern theology, and both have come very far from the black-and-white stances of the late Victorian period. They have long since ceased to draw battle lines. We live in a world where technological detachment is inadequate to solve the problems we face, where the extremes of mathematical rigor begin to strike mathematicians as sterile, and where scientists acknowledge that their most basic principles are themselves tenets of a specialized faith.

If technology and the pursuits of the mind are subjective, they remain willful. But in the medieval view, that willfulness could be harnessed to the greater glory of God. Historian Kenneth Clark paraphrases a moving contemporary description of the construction of Chartres Cathedral, written in 1144. According to this twelfth-century observer:

> When the towers seemed to be rising as if by magic, the faithful harnessed themselves to the carts which were bringing stone, and dragged them from the quarry to the cathedral. The enthusiasm spread throughout France. Men and women came from far away carrying heavy burdens of provisions for the workmen—wine, oil, corn. Among them were lords and ladies, pulling carts with the rest. There was perfect discipline and a most profound silence. All hearts were united and each man forgave his enemies.[1]

But before the construction of cathedrals and all the other miracles of medieval invention went into high gear, another Anglo-Saxon monk gave us a powerful lesson in what it meant for him to be cast in the image of the Master Craftsman. The noted twelfth-century English historian William of Malmesbury recorded this event, which took place just after the year A.D. 1000. He says about the monk Eilmer of Wiltshire Abbey:

> Eilmer...was a man learned for those times,... and in his early youth had hazarded a deed of remarkable boldness. He had by some means, I scarcely know what, fastened wings to his hands and feet so that, mistaking fable for truth, he might fly like Daedalus, and, collecting the breeze on the summit of a tower, he flew for more than the distance of a furlong. But, agitated by the violence of the wind and the swirling of air, as well as by awareness of his rashness, he fell, broke his legs, and was lame ever after. He himself used to say that the cause of his failure was forgetting to put a tail on the back part.[2]

In other words, this rash monk actually achieved a modestly successful glider flight over a distance of more than two football fields. The story gains credence when Eilmer tells us, from across the space of a millennium, that the reason he crashed was that he had failed to equip his glider with the tail it would have needed to give it lateral stability.

We have been raised with the idea that flight was finally achieved only in the lifetime of people who are still living. Yet not only the *dream* of flight but the *fact* of it as well has been with us for thousands of years. Eilmer's flight had its own historical antecedents. Somewhat sketchy accounts make it clear that a successful glider flight was made in the year 875 by a Moorish inventor named Ibn Firnas, who lived in Cordoba, Spain. Firnas also crashed, and he too mentioned his failure to provide a tail. The story of Firnas' attempt had almost surely reached Eilmer. Too bad Eilmer didn't listen to the part about the tail!

Both these bold and imaginative prototypes of the glider were developed in religious and intellectual environments that fostered invention. Ibn Firnas lived at the height of the golden age of Islamic art and science, and Eilmer belonged to the Benedictine order with its tradition that God is the Master Craftsman.

William of Malmesbury, also a Benedictine, undoubtedly knew monks who, as lads, had actually spoken with the aging Eilmer. William writes in the voice of one who delights in Eilmer's bold attempt to harness God's world by natural means. And yet the Benedictine order was no longer the same as it had been in Eilmer's day.

In the year 1000 the Black Friars of the Benedictine order were functioning as a driving force for learning and for technology. But great epochs fade. So the Cistercian monastic order was founded as a reform of the Benedictine order in the year 1098. Saint Bernard took charge of the Cistercians fourteen years later and moved the order in a direction that would complete the transformation of European civilization. The Cistercians were still a branch of the Benedictines, but they were a strict branch that fled worldly commerce to live remote from the habitation of man. Under Saint Bernard they achieved this life by the seemingly contradictory tactic of creating economic independence based on the highest technology of the day.

During the preceding few centuries, the water wheel had revolutionized western Europe by providing a cheap and convenient power source. It had replaced the backbreaking labor required to grind grain, full wool, and saw wood that had been the beginning and end of most people's lives. The Cistercians finally showed how far the water wheel could be taken.

By the middle of the twelfth century their order had reached the cutting edge of hydropower and agricultural technology. A typical Cistercian monastery straddled an artificial stream brought in through a canal. The stream ran through the monastery's shops, living quarters, and refectories, providing power for milling, woodcutting, forging, and olive crushing. It also provided running water for cooking, washing, and bathing, and finally for sewage disposal. These monasteries were, in reality, the best-organized factories the world had ever seen. They were versatile and diversified. Of course, they represented a rather strange way of living remote from the habitation of man, but that is another matter.

We are too often led to see this period of history as a Dark Age. The people who gave us the written record of medieval political history were generally remote from the world of making things. The scribes of the kings wrote about armies and slaughter and had little to say about the engineers who were really changing the world. And the engineers of the Cistercian order did more than just develop this new technology. They also spread it throughout Europe during the twelfth and thirteenth centuries. Their 742 monasteries were major agents of the changes that had completely altered European medieval life by the middle of the thirteenth century.

By A.D. 1200 another power-generating technology took its place alongside the water wheel. The medieval windmill, with its driving fan facing into the wind, was likely an adaptation of the quite different Arab windmill. Arab windmills had vertical fans mounted inside a tower that admitted the wind through vertical slots in its side. The Arabs had been building such windmills for several centuries when the Crusaders arrived.

The powerful European windmill that finally burst upon northern Europe in the last years of the twelfth century typically generated twice the power of a medieval water wheel (five or six horsepower as opposed to two or three). Windmills did not displace water wheels, but they supplemented them in different kinds of environments.

At this point, medieval millwrights had two radically different power sources, yet both ground grain. And we are left with what is, by our standards, an intriguing and difficult question: To the medieval mind, what powered those mills? What invisible efficacy was taken from the water or the wind to grind grain? The wind, in particular, still captures our imagination even though we know about air, kinetic energy, and force balances. What did people see happening in a mill when they didn't have any such knowledge?

A little word game yields a startling insight. The ancient tongues all used the same word for "wind," "breath," and "soul." In Sanskrit that word is *atman*; in Latin it is either *spiritus* or *anima*; in Hebrew it is *rauch*; and in Greek it is *pneuma*. *Rauch* shows up again as the German word for "smoke," and we see pneuma in air-related words such as *pneumatic*. The Russian word for "spirit," *duh*, has many wind-related cognates. *Duhovyia intrumenti*, for example, means "wind instruments."

The connection is that both the wind and the soul were the breath of God. Genesis, for example, begins with God breathing a soul into Adam. Medieval engineers saw nothing less blowing their windmills. The power source was mystical. By the way, some historians are pretty sure that the windmill itself was derived from ancient Buddhist prayer wheels that were spun by sail-like propellers. We can describe the wind in technical terms today, but we haven't let it lose its metaphorical power. In his "Ode to the West Wind," nineteenth-century Romantic poet Percy Bysshe Shelley shouts at the west wind,

Be thou, Spirit fierce,
My spirit! Be thou me, impetuous one!
Drive my dead thoughts over the universe,
Like wither'd leaves, to quicken new birth.

His wind was not just *spiritus;* it was a renewing spirit, a cleansing new broom—the same imagery we use when we tell each other, "It's an ill wind that blows no good."

For medieval engineers, the most plausible explanation of mechanical power was that it *was* the breath of God. And, by the middle of the thirteenth century, the expanded use of water and wind power had created a kind of mania for such power. The power at the disposal of the average person had roughly quadrupled, and when people saw what it could do for them, they wanted more—and more. The solution that arose during the mid-thirteenth century should not really be a great surprise. People looked for means of harnessing what is called in Latin *perpetua mobile*—perpetual motion.

When you and I talk about a perpetual-motion machine we usually mean one that produces power without being fed an even greater amount of power in a different form—say, an engine that produces electrical energy without eating even more energy in the form of coal. Since 1850 we have all agreed on thermodynamic laws that tell us such a machine cannot exist.

But think for a moment like a medieval engineer, hungry for yet more power. For years we have harnessed the motions of wind and water. We have watched our water wheels turn and turn and turn. We have watched our windmills turn, stop for a while, then turn some more. Our eyes tell us that perpetual motion obviously is possible, for the breath of God is always there.

Then, in A.D. 1150, the Hindu mathematician Bhaskara proposed a machine that would produce continuous power. It was simple enough—a wheel with weights mounted around its rim so they swung radially outward on one side and inward on the other. The wheel was supposed to remain out of balance and turn forever. Medieval engineers knew nothing about the conservation of either energy or angular momentum. They had no way of understanding that such a machine was doomed to fail.

The overcentered wheel reached the Moslems in 1200 and France by 1235. For the next five hundred years, writer after writer recommended this ingenious, if impossible, little device. Did they ever try to make one? They did, of course, and the machines always failed miserably. Yet to minds that believed perpetual motion possible, failure merely meant they did not yet have the proportions quite right. The machines all failed, yet that did not keep hope from remaining alive and well.

A late-seventeenth-century form of Fludd's earlier perpetual-motion machines, shown in Böckler's *Theatre of New Machines*. The reservoir provides water power for both the mill and the Archimedean pump that forever resupplies the reservoir.

It was not until the seventeenth century that scientists finally recognized its impossibility. Not until the eighteenth century were mathematical engineers able to show that the overcentered wheel failed to properly conserve momentum. For centuries afterward, each newly revealed physical phenomenon awakened new hopes. Each newly discovered force of nature drove people to look for ways of exploiting it to produce power without consuming energy. The search for perpetual-motion machines finally led contributors to a better understanding of static electricity, surface tension, magnetism, and hydrostatic forces.[3]

The search for perpetual motion has thus been a powerful technological driver, one that remains alive even today. Some people look for it in the face of the physics that says it is impossible, while others simply look for as yet unthought-of ways to keep producing power—for new means of tapping into the breath of God.

The first fruit of the search for perpetual motion was the invention of the long-term repetitive motion of the mechanical clock, which was invented around A.D. 1300—give or take a little. The very person who first promulgated perpetual motion in France was the architect Villard de Honnecourt. After his experiments with the overcentered wheel, Villard carried the theme of perpetual motion forward in a different way. He invented a device that pointed at the sun rising in the east and then kept on pointing toward the sun until it set in the west. That device was the first weight-driven escapement mechanism—the heart of what would soon become the mechanical clock (see Chapter 5). Something like perpetual motion had now been realized in a kind of sustained motion that was entirely feasible.

The mechanical clock calls to mind its antithesis, the hourglass. How old do you suppose the hourglass is? Two thousand years? Four thousand years? It was invented about the same time as the first mechanical clocks, so it is only about seven hundred years old.

The hourglass had some strong characteristics. On the positive side, it was far simpler and cheaper than the mechanical clock or the earlier water clock; resetting it after it ran down couldn't be easier; and it didn't vanish when you used it, the way a graduated candle did. Its accuracy was not bad once some problems had been solved. You could not just load any old sand into it. You had to find a free-flowing material that was unresponsive to humidity.

On the downside, hourglasses were short-term timepieces. The very name suggests it is hard to make one that will run more than an hour. In addition they cannot be calibrated. Sand moves downward in jerks. The edge of the sand is uneven. If you mark five-minute intervals on the glass, the sand will hit those marks differently each time you turn it. An hourglass shows only when an hour is up.

Hourglasses found their niche in setting off blocks of time: the time between canonical hours in a monastery, or between watches on board a ship. They ran neither long enough nor accurately enough for marine navigation. They were a poor person's timepiece—a kind of clock for everyman.[4]

Both the mechanical clock and the hourglass played powerful sym-

A modern hourglass.

bolic roles during the Renaissance. The complex mechanical clock with its rotary gears became a metaphor for the heavenly spheres or the wheel of fortune. But the hourglass, whose sands run out, became a metaphor for the destiny we all inevitably face, and has remained a universal symbol of death.

Two technologies, one simple, one complex, running side by side—the clock making a continuum of time and the hourglass segmenting it, the clock speaking of timelessness and the hourglass showing us finality, the clock evoking things celestial and the hourglass reminding us of base earth. Two technologies, yin and yang. Why was the hourglass so late in coming? Maybe it had to wait for its antithesis, the mechanical clock, to be invented.

Two hundred and fifty years after their invention, mechanical clocks had become very sophisticated machines that, in retrospect, provide insight into the nature of invention. As machines go, clocks have their own distinctive character. You wind them up and then sit back to watch them carry out their function. A well-designed clock goes on and on, showing the time of day without human intervention and without self-correction. That is exactly why the ideal clock—the clock that we almost but never quite make—became a metaphor for divine perfection.

By the middle of the sixteenth century, clocks had not only become passably accurate. They had also become remarkably beautiful, adorned with stunning but seemingly useless trimming. Mechanical figures marched out on the hour and performed short plays. Extra dials displayed the movements of planets. Clocks were crowned with exquisite miniature gold, bronze, and silver statuary.

The intricate wheels and gears of these baroque clocks became a metaphor for the solar system, the universe, and the human mind, as well as the perfection of God. The best minds and talents were drawn into the seemingly decorative work of clock making because clocks harnessed the imagination of sixteenth-century Europe. All this was rather strange, because there was little need for precision timekeeping. Later, during the eighteenth century, the clock began to assume its role as a scientific instrument—especially for celestial navigation. But in 1600 the search for accuracy was primarily an aesthetic and intellectual exercise.

Our thinking is so practical today. You or I might very well look back at the sixteenth century and condemn elegance in clock making as a misuse of resources. But the stimulus of the clock eventually resulted in previously unimagined levels of quality in instrument making. It drove and focused philosophical thinking. In the end, the precision of this seemingly frivolous high technology was a cornerstone for the seventeenth-century scientific revolution, for eighteenth-century rationalism, and, in the long run, for the industrial and political revolutions that ushered in the nineteenth century.

Sixteenth- and seventeenth-century clock making was the work of technologists who danced to their own freewheeling imaginations and aesthetics—technologists who were having fun. Technologists like that really change their world. And make no mistake, baroque clock makers set great change in motion. In that sense the mechanical clock marked the end of the medieval era. It was more a creature of the age that followed (more about that in a moment).

The defining technology of the high Middle Ages was not the agitated moving clock, but the static Gothic cathedral. It is hard to say too much about Gothic cathedrals. They combine immensity with a delicacy of balance and detail that must be seen to be believed. The spire of Strasbourg Cathedral, for example, is almost as high as the Washington Monument. Gothic architecture appeared suddenly in the twelfth century and kept evolving for 250 years. Then it abruptly stopped developing toward the end of the fourteenth century.

Rouen Cathedral

The people who created this art appear to have been without formal education. Estimates suggest that only some 40 percent of the master masons could even write their name on a document. They probably knew nothing of formal geometry, and it is unlikely they made any calculations. Most astonishing of all, they built *without working drawings*. The medieval cathedral builder learned his empirical art (or should we call it empirical science?) through apprenticeship. The master builder had all kinds of tricks of the trade at his disposal, many of them jealously guarded. These tricks amounted to a vast inventory of knowledge of material selection, personnel management, geometrical proportioning, load distribution, and architectural design, coupled with a firm sense of liturgy and Christian tradition.

These builders saw no clear boundary between things material and things spiritual. Their art flowed from their right brain. It was visual and spatial. They levitated tons of stone in the air to communicate their praise of God, and when they were finished, they embellished the nooks and crannies and high aeries of their buildings with the phantoms of their minds—with cherubs and gargoyles and wild caricatures of one another.

Gargoyle on the Sainte-Chapelle in Paris.

By the thirteenth century, they boldly and proudly identified themselves with their work. An inscription, twenty-five feet long, on the outside of the south transept of Notre Dame Cathedral, the one pointed toward the Seine River, says:

> Master Jean de Chelles commenced this work for the Glory of the Mother of Christ on the second of the Ides of the month of February, 1258.

So what became of this marvelous art? The best guess is that it died when the master builder became an educated gentleman—when he moved into an office and managed the work of others at a distance. At that point the kind of hands-on creativity that had driven it so powerfully dried up.

Still, the last great medieval cathedral was only recently completed—here in the United States. The National Cathedral in Washington, D.C., was built with great fidelity to both the style and the working esprit of the medieval art. This huge structure was finished after eighty-two years of construction, and it is breathtaking. If you ever visit Washington, don't miss the chance to see it.

Eventually the Church-driven medieval world, with all its high technology, ran to the end of its tether. Overpopulation took its toll. Famines began visiting Europe in the late thirteenth century. Then, in 1347, the plague, caused by the bacterium *Yersinia pestis*, swept out of Asia into a weakened Europe. Rats carried the disease off ships in

Genoa. In just four years it killed a third or so of the people in Europe. It took three forms: *Bubonic* plague hit the lymph system, *pneumonic* plague attacked the lungs, and *septicemic* plague assaulted the blood. But the words *Black Death* encompassed it all.

After 1351, the Black Death went from its epidemic phase, where the disease suddenly appeared, to its pandemic phase, during which it settled into the local environment and kept coming back every few years to whittle away at the population. From the first famines in 1290 until the plague pandemic began to recede in 1430, Europe lost over half its people. And that was after the disease had devastated the Orient. The Black Death is probably the greatest calamity our species ever suffered.[5]

What did the plague leave in its wake? For one thing, it unraveled the feudal system. It also left many survivors wealthy with the material goods and lands of those who had died. Manual labor became precious. Wages skyrocketed, and work took on a manic quality. When death rides on your back, time also becomes precious. It matters. The Church-centered world before the plague had been oddly timeless. Now people worked long hours, chasing capital gain, in a life that could end at any moment. The first new technologies of the plague years were mechanical clocks and hourglasses.

Medicine had been the work of the Church before the plague. Physicians were well-paid, highly respected scholars. They spun dialectic arguments far away from unwholesome sick people—not unlike some of today's specialists. Fourteenth-century medicine, like the fourteenth-century Church, had failed miserably in coping with the plague. Now both medical and religious practice shifted toward the laity. Medicine was redirected into experimentation and practical pharmacology. Medical and botanical books began to appear at the end of the fifteenth century, written not in Latin but in the vernacular, and by a whole new breed of people.

Technology had to become less labor-intensive; it had to become high-tech. For good or for evil, the plague years gave us crossbows, new medical ideas, guns, clocks, eyeglasses, and a new craving for general knowledge. So the rainbow at the end of this terrible storm, all intermingled with the Hundred Years' War, yielded its pot of gold. The last new technology of this ghastly century and a half was the printing press. The new presses finally thawed the epoch that Shakespeare named the "winter of our discontent." They provided access to knowledge, and they started the rebirth of human energy and hope in Europe.

But the brief bright period in which religion and technology had wed was now fragmented. From now on religion and our direct knowledge of the world around us would coexist under an increasingly uneasy peace. Take one example: Down through the centuries, many biblical scholars have tried to calculate the age of Earth. By the mid-nineteenth century a hundred or so such calculations were extant. They gave ages that ranged from fifty-four hundred to almost nine thousand years.

Well before Darwin, geologists had begun insisting that the Bible was not really meant to give us this sort of technical data—that Earth is, in fact, much older. At first they had no basis for making their own estimates. Then, in 1862, the English scientist Lord Kelvin made the first numerical calculation of Earth's age based on data gathered outside the Bible. Kelvin knew Earth's temperature increased one degree Fahrenheit for each fifty feet you went into the ground. He guessed that Earth began as molten rock at seven thousand degrees Fahrenheit. Calculating how long Earth would have to have cooled to establish that surface temperature gradient, Kelvin found that it must have taken a hundred million years for Earth's temperature to level out at one degree every fifty feet.[6]

From the viewpoint of contemporary scientists, Kelvin had opened himself up to assault when he used Fourier's equation to calculate transient heat conduction through the Earth. In 1807 Joseph Fourier had developed an equation for heat conduction that was based on avant-garde mathematics. Kelvin had to invent radical mathematical means to solve Fourier's equation since he did not know conditions deep inside the Earth.[7]

So now the fat was in the fire! The deeply religious and antievolutionist Kelvin had determined an age that was far too young to satisfy geologists and Darwinists. But it was plenty old enough to waken the ire of biblical literalists.

The real problem with Kelvin's estimate was that he did not know about radioactivity. Today we know that Earth's temperature variation is sustained by radioactive decay. That means Kelvin's cooling calculation could not possibly have given Earth's age correctly. Its real value lay in the intellectual stimulus it created. Of course his critics had no more knowledge of radioactivity than he did; the great Victorian scientists and mathematicians knew *something* was wrong, but it was unclear *what* was wrong. So they formed ranks to fight about questions of mathematical method and biblical exegesis. The debate went on until the twentieth century. It drew in Darwin, Huxley, Heaviside, and many more. When they were through fighting, at least mathematical heat conduction analysis had found a solid footing.

Today, modern chemical analysis tells us that Earth is four and a half billion years old. But the debate over Kelvin's calculation helped to set up techniques by which engineers can solve far nastier heat flow problems than he ever could—techniques for determining everything from how long it takes to refrigerate fruit to a certain temperature to how to cool a brake shoe. I do not incline to be dismayed by this checkered story, for it becomes clear that we learn so much more when the path to understanding leads through briar patches like this one.

It took historian Henry Adams to stand away from the nineteenth-century science-religion quandary and put things in perspective by laying them out once more against the backdrop of the medieval Church. Our second president, John Adams, had sired a dynasty. His son John Quincy was also president, and his grandson Charles was a congressman and ambassador to England during the Civil War. In his autobiography, *The Education of Henry Adams*, John Adams' great-grandson measures himself against his extraordinary forebears and finds himself wanting—even though he was a writer, congressman, and noted historian.

Toward the end of the autobiography Adams portrays himself as a sort of everyman facing the juggernaut of twentieth-century science and technology. His chapter titled "The Dynamo and the Virgin" takes us through the great Paris Exposition of 1900. Adams was one of forty million people who visited its eighty thousand exhibits. He was drawn back day after day, trying to understand it all.

Four generations of the Adams family.

Adams' most important works of history up to that point had been studies of two medieval edifices: the abbey at Mont-Saint-Michel and Chartres Cathedral. They had led him to see the remarkable social impact of medieval Christianity, centered as it was on the Virgin Mary.

Now he gazed at dynamos, telephones, automobiles—wholly new technologies that had sprung into being in just a few years and which were based on invisible new forces of radiation and electric fields. He saw that the dynamo would shake Western civilization just as surely as the Virgin had changed it eight hundred years before. His historical hindsight made him comfortable with the twelfth century, but the Paris Exposition was more than he could digest.

His guide through much of the exposition was the aeronautical pioneer Samuel Pierpoint Langley. Langley was a down-to-earth physicist, willing to explain things in functional terms. But Adams was too intelligent to mistake the explanation of how something works for a true understanding of it. He says: "[I found myself] lying in the Gallery of Machines—my historical neck broken by the sudden irruption of forces totally new." Adams didn't mention two ideas that would complete the intellectual devastation before his autobiography was published: quantum mechanics and relativity theory. He probably did not know about either one, but, on a visceral level, he saw them thundering down the road. It was clear enough to Adams as he surveyed the exhibit that it portended a great unraveling of nineteenth-century confidence in science.

In the end, Adams lamented the blind spot of his times, the denial of mystery. The Virgin was the mystery that drove the medieval technological revolution. He said,

> Symbol or energy, the Virgin had acted as the greatest force the Western world ever felt, and had drawn man's activities to herself more strongly than any other power, natural or supernatural, had ever done.

Adams realized that the dynamo and modern science were ultimately being shaped by forces no less mysterious. Neither Langley nor anyone else *understood* radium and electricity, any more than Adams himself did.[8]

History has proven Adams right. Nineteenth-century science and technology came out of the exhibition hall changed beyond recognition. The self-sufficient determinism of Victorian science was breaking up under his nose. By the time the implications of quantum mechanics were sorted out, science was as incomplete as ever. It was a chastened intellectual enterprise that left both the exhibit hall and the nineteenth century. We had come to understand that the Bible is not a book of scientific formulas. But during the twentieth century, we have also come to understand that science too is a glass that lets us only glimpse the mysteries of our being, and then only in part—only darkly.

3

Looking Inside
the Inventive Mind

An inventor—any creative person—knows to look under the surface of what things *seem* to be, to learn what they *are*. I have been able to find only one constant in the creative mind. It is that surprise is the hidden face of the coin of invention. In their operetta *Pinafore*, Gilbert and Sullivan warn us:

Things are seldom what they seem,
Skim milk masquerades as cream;
Highlows pass as patent leathers;
Jackdaws strut in peacock's feathers.

For example, an engineer designing a highway system wants to include crossroads between the major arteries. Common sense says that crossroads will increase driver options and speed traffic. Only very keen insight, or a complex computer analysis, reveals that crossroads tend to make matters worse. They often create localized traffic jams where none would otherwise occur.

We are caught off guard when common sense fails us. Yet it is clear we would live in a deadly dull world if common sense alone were sufficient to lead us through all the mazes around us. If what we learn is no more than what we expect to learn, then we have learned nothing at all. Sooner or later, every student of heat flow is startled to find out that insulation on a small pipe can sometimes increase heat loss. Common sense is the center of gravity we return to after our flights of fancy. But it is the delicious surprise—the idea that precedes expectation—that makes science, technology, and invention such a delight.[1]

A wonderful old expression calls creativity "a fine madness," and it is. Invention lies outside the common ways and means. If it is sane to respond predictably to reality, then invention surely *is* madness. A well-known riddle shows us something of the way that madness works. You are asked to connect nine dots, in a square array, with four straight lines. Each line has to continue from the end of the last line.

The problem seems to have no solution. If, for example, you draw a sequence of lines on three sides, like this, then the fourth line would be either a diagonal that connects the center dot, or a horizontal line that connects the lower dot. You cannot get them both. The trick, of course, is to walk around accepted limitations. It is easy enough to connect the nine dots when we work without knowing the answer ahead of time. Once we recognize that lines need not conform to the square space suggested by the dots, we can solve the problem. The expression "thinking outside the box" probably comes from just this riddle. (Turn the page for the solution.) What kind of mind does it take to see through a question like this? Certainly not one that thinks "normally"!

That's why creativity cannot be reduced to method. The best we can do is to meet a few of the great inventive minds and inventive acts. When we do that, we find some creative people adopting a camouflage of orthodoxy—focusing their creative power productively and living contentedly. Others have been savaged by the fine madness running amok in their lives.

So let us flirt with this kind of madness. Let us gaze on some things that are not what they first seem to be. Let us meet some creative people and their acts of invention. Let us savor the delicious surprise that made it worth the cost of their travels through an alien land. Let us begin with the story of Christopher Wren and the dome of Saint Paul's Cathedral.

When I was young in St. Paul, Minnesota, I would take the streetcar downtown. It would rattle past a great cathedral with a lovely big dome. The name of that cathedral was both the name of the city and the name of the great English cathedral that it copied—St. Paul's in London. No child of World War II can forget movies showing St. Paul's dome standing out against London's bomb-lit night sky. It survived the blitz even though it had been rebuilt repeatedly since A.D. 607.

In 1665 it was a five-hundred-year old Gothic cathedral, never really finished and falling apart. The brilliant young architect Christopher Wren was told to rebuild the old wreck. He put a radical plan before the cathedral planning commission: He would tear down the old Gothic building and replace it, capping the new building with a dome like the ones on Renaissance churches in Europe. The commission would have none of that. Cathedrals had spires, not domes. St. Paul's would be patched, not rebuilt, and Wren would place a new spire upon it.

Wren rankled for a year. Then nature intervened. The terrible Great Fire of 1666 finished off the old building, and Wren was free to design a new one. The commission still rejected his dome, even though King Charles rather liked it. Finally the king gave Wren a loophole. He told him to erase the dome from his plans and draw in a steeple—any steeple. Then the king put a phrase in Wren's contract that granted him "liberty to make such variations as from time to time he should see proper."

Wren topped his design with a hideous, out-of-proportion steeple, satisfied the commission, and went to work. It took thirty-five years to build the new cathedral—far longer than the collective memory of any committee. As the structure rose, Wren made a careful sequence of changes. The bogus steeple gracefully gave way to a marvel of engineering: a great dome, 110 feet across, soaring 368 feet into the air. To hold it without buttresses, Wren girdled it with a huge iron chain hidden by the facing stone. (That old, rusted chain was finally replaced with stainless steel in 1925.)

Today you can still go to London, sit in a pew, and see what ninety-two-year-old Christopher Wren saw on the last Sunday of his life—an interior that seemingly stretches to infinity in five directions, north, south, east, west, and up. The view upward, past the whispering gallery to that splendid ceiling, is hard to forget. When Wren died, they buried him in the cathedral under a small plaque. The building itself is his monument. It is also a monument to the will of this gentle genius who found a way to show people what could not be explained to them.[2]

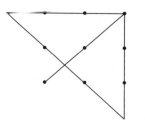

So we begin with an exemplary case—a man who, like his medieval forebears, found fine constructive means for living in the world of rough-and-tumble with his creative daemon at his side. We find similarities in the story of Benjamin Thompson, but Thompson's means for coping with the daemon reveal harder edges than Wren's did.

Thompson was raised in Woburn, Massachusetts, in the years of the gathering American Revolution. He wrestled out a homemade education in Boston and, when he was only eighteen, went off to Rumford, Massachusetts, as the new schoolmaster. He soon married a wealthy thirty-one-year-old widow. Then he took up spying on the colonies for the British. When the colonists found out what he was doing, he deserted his wife and baby daughter and fled to England. Thompson devoted the next several years to shameless social climbing that eventually put him in a high-ranking position with the Bavarian court in Munich.

Here his life took on a different coloration. He boldly combined technical insight with social reforms that were years before their time. He instituted public works, military reforms, and poorhouses. He equipped public buildings with radical kitchen, heating, and lighting systems. As a part of a lifelong interest in heat and its management, he gave us the Rumford stove. And why Rumford? Therein hangs the rest of his story.

In 1792 he was made a count of the Holy Roman Empire and he took the name of the town of Rumford. Thompson is best remembered as Count Rumford and for the experiments he made under that name five years later. His interest in field artillery led him to study both the boring and the firing of cannons. There he saw that mechanical power could be converted to heating, and that there was a direct equivalence between thermal energy and mechanical work.

People at the time thought that heat energy was a fluid—a kind of ether called *caloric*—that flowed in and out of materials. Caloric could not be created by mechanical work or by any other means. Rumford's results flew in the teeth of the caloric theory. They showed that you can go right on creating caloric as long as you have a source of mechanical work. He also showed that when you fired cannons without a cannon-ball in the barrel, they heated up more rapidly than when you fired them loaded. When the cannonball was there, heat was converted to the work of accelerating it.

Rumford's story eventually took a last, ironic turn. Caloric had been given its name by the famous French chemist Lavoisier, who was beheaded during the French Revolution. When Rumford returned to England and France, he became involved in a four-year relationship with Lavoisier's widow. Their affair ended in a disastrous and short-lived marriage. Before the marriage Rumford crowed, "I think I shall live to drive caloric off the stage as the late Lavoisier 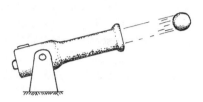 drove away phlogiston. What a singular destiny for the wife of two Philosophers!!" Count Rumford was indeed instrumental in driving caloric off the stage. But can it be any surprise that the marriage failed?[3]

Rumford clearly had at least some dimension of control over the daemon. But many people have none. A young man born in France in 1811, three years before Rumford died in France, is a fine case in point. He was Evariste Galois, the father of modern algebra. Galois died of gunshot wounds at the age of twenty years and seven months. He was still a minor when his brief, turbulent life ended.

He began a career in mathematics by twice failing the entry exam for the Ecole Polytechnique because his answers were so odd. He was finally accepted by the Ecole Normale, only to be expelled when he attacked the director in a letter to the newspapers. A few months later he was arrested for making a threatening speech against the king. He was acquitted, but he landed right back in jail when he illegally wore a uniform and carried weapons. He spent eight months there, writing mathematics. As soon as he got out, he was devastated by an unhappy love affair. It might be fair to say he was a typical bright young teenager.

For some murky reason—perhaps it was underhanded police work—he was challenged to a duel on May 30, 1832. It was a duel he could not win, but which he could not dodge either. By then his talents as a mathematician were known. He had published some material, and outstanding mathematicians including Gauss, Jacobi, Fourier, and Cauchy knew of him.

On May 29 he wrote a long cover explaining the hundred or so pages that he produced during his entire short life. He set down what proved to be the foundations of modern algebra and group theory. Some of his theorems were not proven for a century. He faced death with a cool desperation, reaching down inside himself and getting at truths we do not know how he found.

His fright and arrogance were mixed. The letter was peppered with asides. On one hand, he wrote: "I do not say to anyone that I owe to his counsel or…encouragement [what] is good in this work." On the other hand, he penned in the margins, "I have no time!" When poet Carol Christopher Drake heard his story in 1956 she wrote:

Until the sun I have no time
    But the flash of thought is like the sun
    Sudden, absolute:

    watch at the desk
    Through the window raised on the flawless dark,
    The hand that trembles in the light,
    Lucid, sudden.

Until the sun
I have no time
    The image is swift,
    Without recall, but the mind holds
    To the form of thought, its shape of sense
    Coherent to an unknown time—

I have no time and wholly my risk
Is out of time; I have no time,
I cry to you I have no time—
    Watch. This light is like the sun
    Illumining grass, seacoast, this death—

I have no time. Be thou my time.[4]

The next morning Galois was shot, and two days later he was dead; his creative daemon had killed him. Still, he had done more for his world in one night than most of us will do in a lifetime, because he knew he could find something in those few moments when he had to reach down and look deeply inside himself.

Mathematics, of course, is terribly demanding. To do mathematics and to do it well requires enormous mental discipline. Some have the technical discipline without having the supporting emotional self-control. That fact came home to me the day a colleague pressed me to write a radio program about Georg Cantor.

I began looking closely at Cantor and was immediately reminded of a *Time* magazine article my father had read to me when I was in grade school. The article snatched my imagination. Someone proposed a new number called the *googol*. The googol was ten to the hundredth power—a one followed by a hundred zeros. Later I learned that it would be of little help in counting real objects, because we are hard pressed to find that many real objects in the whole universe—even atoms.

Still, where does counting end and infinity begin? There is good reason to ask about infinity. Every engineering student knows that infinity is not just the end of numbers. If we ask how real systems behave when velocities, time, or force becomes infinite—if we ask about the character of infinity—we get some unexpected answers.

The mathematician Georg Cantor also wondered about infinity. He was born in Russia in 1845. He wanted to become a violinist like his mother, but his father, a worldly merchant, wouldn't have it. When he was seventeen, his father died. Cantor went on to finish a doctorate in math in Berlin when he was only twenty-two. His career was not long—he burned out before he was forty and spent the rest of his life in and out of mental illness.

But what he did was spectacularly important, and it arose out of an innocent counting question. He began with an idea we find even in Mother Goose. Remember the rhyme "One potato, two potato, three potato, four. Five potato, six potato, seven potato, more?" Counting is like matching one set of things with another—in this case, numbers with potatoes. Cantor asked if counting all the infinite number of points on a line was like counting all the infinite number of points in a surface. To answer the question, he had to invent something called *transfinite numbers*. He identified orders of infinity. If you simply count the integers, you obtain the first order of infinity. Raise two to that power and you get the next order of infinity, which happens to be the number of sets you can form from the lowest order of infinity, and so forth. To sort through all that, he had to invent set theory, which has become a basic building block of modern mathematics.

Cantor fell into an odyssey of the mind—a journey through a strange land. He had to overcome the resistance of his father, of the great mathematicians, and even of his own doubts. When he was thirty-three he wrote: "The essence of mathematics is freedom." To do what he did, he had to value freedom very highly—freedom coupled with iron discipline, freedom expressed through the driving curiosity of a bright child, freedom to pursue innocent fascination until that fascination finally changed our world. Cantor lived his troubled life until 1918—long enough for him to see set theory accepted and himself vindicated for his soul-scarring voyage of the mind.[5]

But Cantor's and Galois' cases are extreme. While stories like theirs abound, the telling of them can lead us away from the better aspects of the creative process. Let me offer a third mathematician/engineer/scientist whose story has a completely different flavor—or perhaps I should say a peculiar lack of any flavor at all. Let us meet Josiah Willard Gibbs.

Historians dislike superlatives. It is too easy to be wrong when you use words like *first* and *best*. Yet I am completely comfortable saying that Gibbs was the greatest American scientist who has ever lived. Gibbs was born in New Haven, Connecticut, in 1839. He was the seventh in an unbroken line of American academics, stretching all the way back to the seventeenth century. His father was a noted professor of linguistics at Yale University. He lived his entire life in the same house and died there in 1903.

What did Gibbs do? He created the entire subject of chemical thermodynamics; he gave us the subject of vector analysis; he invented statistical mechanics and developed it as far as it would go before quantum mechanics could take it further; and he did basic work in several other areas. Other great scientists contribute to fields. Gibbs *created* three entire fields—pulled them out of his empyrean mind and gave them life.

He did little to invite fame. He hardly traveled; he did not collaborate; he never married; and he published most of what he wrote in the obscure *Transactions of the Connecticut Philosophical Society*. Outwardly, he was dry and colorless.

Gibbs studied at Yale, where he earned one of the first doctorates offered in the United States. He was America's very first Ph.D. in mechanical engineering. His thesis dealt with shaping gear teeth. After that, he taught at Yale and worked on the design of railway equipment, particularly brake systems. From 1866 to 1869 he studied mathematics and physics in Paris, Berlin, and Heidelberg. That was the only real trip he ever took. He was thirty-four before he published his first paper, and it was still later that his abilities started to become apparent. When his work on thermodynamics attracted attention, the Johns Hopkins University offered him a position. Up to then, Yale had not been paying him. Now, at least, Yale put him on the payroll.

Gibbs' work is spatial—like good engineering work. It moves in a surrealistic, multidimensional landscape. People who join him in his voyage of the mind find it seductively beautiful. Since his death, twentieth-century scientists have peeled him like an onion, and under each layer they have found what they had missed under the previous one.

J. Willard Gibbs' life was wrapped in plain gray—faculty meetings, committees, classes. He lived as a quiet professor doing commonplace things. He received no major grants, no Nobel Prize. But it was his edifice that Einstein and Fermi completed. He rewrote science. He changed history. And I can assert with a bland certainty that he was our greatest scientist.[6]

Gibbs is a curious case in that he offers almost no guidance whatsoever as to how you and I might reach for the brass ring of creativity. I can only lay him out before you and say, "Make of him what you will." For guidance, we must look elsewhere. So let us return to the remark we made at the outset—the observation that creativity is surprise. This means seeing ideas out of context. The creative process must include an ability to recognize the stray idea out on the far fringe of our peripheral vision.

Consider that our knowledge of microbes is just over three hundred years old. They were first observed in the late seventeenth century by the Dutch lens maker Antonie Van Leeuwenhoek. He called these small creatures *animalcules*, and he realized that they swam in any body of water. But what about relating these small beasts to disease? One hundred and fifty years later, disease was rampant in London. Half the newborn babies died, and the death rate was far higher among the poor than among the wealthy. Two things had become clear by then. One was that London's drinking water was, by and large, loaded with microorganisms. The other was that filth—particularly raw sewage—was to be found everywhere in poor areas.

It is obvious to us that germs were causing the diseases. But germs, after all, swam in waters drunk by both the well and the sick. It was obvious that bad smells were found in unhealthy neighborhoods, and it seemed clear that stench, or *miasma*, as it was called, caused disease. No one had cause to worry about the water. It was the stink that people felt they had to eliminate.

Then, in 1849 and 1853, London suffered terrible epidemics of cholera. In 1853 a physician named John Snow began looking at statistics. He worked doggedly among the sick and kept meticulous records of who had died and exactly where they had lived when they fell ill. It took a long time, but Snow eventually found a high incidence of cholera among people who had been drawing water from a source called the Broad Street well.

That made no particular sense until Snow realized that the cesspool of a tenement occupied by a cholera patient was adjacent to the well. Contaminated water had been leaking from the cesspool into the groundwater picked up by the well. Over protests, he managed to remove the handle from the well, and cholera abated in that part of London.[7]

Snow's report soon led people to see that cholera was caused not by noxious gases but by what came to be called *fecalized water*. He put people on the track of the real disease agent. Four years later, Louis Pasteur connected disease to bacteria. And in 1865 Joseph Lister found that he could kill disease-carrying bacteria during surgery by spraying a carbolic acid solution over the patient. Finally, in 1882, a scant twenty-nine years after Snow pinpointed the Broad Street well, the German physician Robert Koch showed how to make a disease-specific vaccine. He had found the bacterium that caused anthrax and figured out how to make a vaccine to prevent it from infecting animals.

It can take decades for people to overturn old thinking. The leap from unhealthy vapors to bacteria was still a hard one to make, even once Snow had used the Broad Street well to show that a leap had to be made. Making that leap, of course, reminds us that no accounting of the creative process is complete without an eccentric-scientist story, even if such tales miss the point. Creativity, after all, is always eccentricity. It can be no other way. But in the interests of completeness, let us meet the eccentric Nikola Tesla.

When I worked in Yugoslavia back in 1974, I especially liked their five-hundred-dinar notes. In that economy, they were a kind of thirty-dollar bill with a picture of a lean man reading from a large book. The man was Nikola Tesla—one of the early geniuses of electricity. Tesla was born in Serbia and educated in Prague. He came to the United States in 1884, when he was twenty-eight years old.

Looking Inside the Inventive Mind

Tesla had by then revealed a wild mercurial talent for manipulating the mysterious new forces of electricity, and he carried a letter of introduction to Thomas Edison. The already famous Edison would probably have shrugged him off, but he was shorthanded when Tesla showed up. The new electric lighting system in the steamship *Oregon* was failing, so he hired Tesla on the spot and sent him off to fix it. Tesla did fix it, but he lasted less than a year with Edison. Edison took Tesla for an egghead, a theoretician, and an epicurean snob. Edison said that 90 percent of genius was knowing what would not work. Tesla called that kind of thinking an "empirical dragnet." He complained,

> If Edison had a needle to find in a haystack he would proceed at once with the diligence of a bee to examine straw after straw until he found [it]. I was a sorry witness of such doings,… a little theory…would have saved him ninety percent of his labor.[8]

Tesla was a dapper bon vivant, six and a half feet tall. He spent every cent on the good life. He cultivated rich and famous friends, people such as Mark Twain. He wrote poetry and spoke half the languages of Europe. But he never married. In fact, he could not bear physical contact with other people because he had a terrible phobia about germs.

He eventually found his way to George Westinghouse, to whom he gave the concept of alternating current. That led to open combat with Edison, who clung to direct current long after it was clear he was riding the wrong horse. But Tesla was not just the father of the alternating-current power that serves your home. He also demonstrated the concept of the radio before Marconi did. He invented the "Tesla coil," a resonant transformer that generates spectacular million-volt sparks. He was the electrical showman of the late nineteenth century, dazzling audiences with brilliant electrical displays.

The unit of magnetic-flux density was named after Tesla, yet he never published a technical paper. Lord Rayleigh once told him to settle down and specialize like a proper scientist. That was poor advice for the wild Serbian cowboy who rode behind Tesla's urbane front. Tesla played counterpoint to Rayleigh's orthodoxy just as he did to Edison's dogged trial-and-error methods. Edison and Rayleigh were great men indeed, but Tesla reminds us once more that there is no formula for calling up the muse of invention. It is something that ultimately might be impossible to teach.

Mark Twain (Samuel Clemens) doing a high-voltage experiment in Nikola Tesla's laboratory, from *Century Magazine*, April 1895.

So what of the people who attempt to teach engineering? How do they awaken the creative genie? Well, sometimes they do and sometimes they don't. Let us meet two men who were exemplars in very different ways. What was it that they did right?

First I give you Max Jakob. In 1937 Jakob and his family boarded the steamer *Berengaria* in Cherbourg, France, to make a stormy six-day crossing to New York. The Jakobs were leaving their German homeland for good. Max Jakob limped slightly as he walked up the gangway, the result of a wound on the Russian front in World War I. He was now fifty-eight years old and a prominent German expert in the field of heat transfer.

For thirty-one years, ever since he had finished his doctorate in Munich in 1906, Jakob had worked on central questions of the thermal sciences, and he had been involved with the greatest scientific minds of his era. His daughter, Elizabeth, showed me some of the remarkable items from his correspondence. A postcard from Einstein (who was born and died within months of Jakob) indicates his gratitude to Jakob for setting a critic straight on relativity. A letter from Max Planck thanks Jakob for correcting an error in his paper.

But now Jakob had to board a boat to America. Four years before, German troops had gone through Berlin painting the word *JEW* in large white letters on store windows; Jakob wrote in his diary, "I never valued my Jewishness as much; but today I'm happy not to be on the side of those who tolerate this." First he sent Elizabeth to college in Paris. Then he and his wife prepared to leave what seemed the intellectual and cultural center of the universe. They were moving to the land of gangsters and Al Capone—to Chicago, whose climate, he had been told, was murderous. Each of them was allowed to take four dollars out of Germany.

Jakob's move proved terribly important for America. We were far behind Germany in understanding heat flow and were working hard to make up ground. Jakob gave us the first direct conduit to that knowledge. His students formed a cadre that practically defined the field from the 1950s through the '70s. And America was a pleasant surprise for Jakob. The first photographs show him smiling and inhaling the fresh air of a new land. Once here he discovered, and took pleasure in, civilization of a different form, but civilization nevertheless. He found youthful excitement in the intellectual climate, and he became a part of it. He gave every bit as much as he got. Many of today's elder statesmen in the field were his students, and everyone in the field knows his work.

In a diary entry the week before his death, in 1954, he writes about hearing the great contralto Marian Anderson sing. He was deeply moved by her singing of "the magnificent Negro spirituals, especially 'Nobody Knows the Trouble I've Seen' and 'He's got the Whole World in His Hands.' " By then America had claimed its day as the world leader in heat transfer. Jakob and his many students had helped us so much to get there. He finished his life as one of the great *American* engineers and educators.

The story of Llewellyn Michael Kraus Boelter has a completely different flavor than that of Max Jakob. Only the qualities of steadiness and evenhandedness bind the two together. Both men worked primarily on problems of heat transfer, making use of those physical laws that govern the flow of heat.

Those laws were known by the beginning of this century, but there is a great gulf between knowing raw physical laws and knowing how to make them serve a world that is hungry for energy. In 1900 it was practically beyond us to calculate heat flow in most situations. Then German engineers began to develop the mathematical means for using these laws. By the 1920s they had made great strides, and America, by comparison, was far behind. That all changed by World War II, and the person who contributed the most to this change before Max Jakob moved here was Boelter.

Boelter was born in 1898 and raised on a Minnesota farm. He graduated from the University of California at Berkeley and then stayed on as a faculty member. There he gave us all a lesson in how to teach engineers.

His method was simple. He went directly to the student's mind. Nothing got in the way, least of all himself. He taught students to go to the unanswered question—to attack their own ignorance. He directed their efforts into the study of heat transfer. He drew people in, and together they opened up the German literature. Then they kept right on moving while the German work was being ended by Nazi repression. By the early 1930s they had created a new American school of heat transfer analysis.

Knowledge was the great equalizer for Boelter. He and the people around him used last names—no titles. Even Berkeley's catalog listed everyone as Mr. The unanswered question was the only absolute taskmaster, and he made sure his students faced questions that led them where they had not been before. He saw to it that it was his students, more than he, who went on to become the international leaders in the field.

He went to UCLA as dean of engineering in 1944, and he made a great learning laboratory of the place. He abolished departments. Were you an electrical engineering professor? Fine, this term you would teach a civil engineering course. That way knowledge stayed fresh—learning stayed alive. In 1963 he spoke to the UCLA freshmen. His words look repetitive on the written page, but then you catch their antiphonal rhythm—with the lines alternating the first word *you* and *and*. Coming from this quiet man, they have an astonishing intensity and an unexpected moral force. He said:

> The products of your mind are the most precious things you own, that you possess. And you must protect them, and must not do wrong with them, you must do the right thing. You must always have in mind that the products of your mind can be used by other people either for good or for evil, and that you have a responsibility that they be used for good, you see. You can't avoid this responsibility, unless you decide to become an intellectual slave, and let someone else make all of these value judgments for you. And this is not consonant with our democratic system in this country. You must accept the responsibility yourself, for yourself, and for others.[9]

Boelter really saw the transcendence of thought—of the "products of the mind." He saw their power to change life. He understood that they bind us to responsibility, but that they also set us free. Boelter was, in essence, giving his students advice on how to control the creative daemon who sits at our side when we function creatively.

Now let's look at another kind of case history—one about the creation of the daemon itself. It is a story that began as the summer of 1896 drew to a close. Orville Wright had caught typhoid fever, and his older brother, Wilbur, sat at his side, nursing him as he hovered at death's door for six weeks. Sixteen years later Wilbur caught typhoid, and it

Looking Inside the Inventive Mind

killed him. In between, of course, the two brothers created the first workable airplane.

Before Orville fell ill, newspaper articles about Otto Lilienthal, the German pioneer of flight, had deeply impressed the brothers. Lilienthal built and flew gliders until he died in a crash in 1896, the same year Orville almost died. Later, Orville claimed that Wilbur read about Lilienthal's death while he was ill and had withheld the terrible news until he recovered. The story may have been distorted by time, as Lilienthal actually died a month before Orville fell ill. What it really tells us is that the two powerful events of Lilienthal's death and Orville's recovery were linked in the Wright brothers' minds.

Lilienthal built gliders for six years. Other people had made gliders before he did, but no one had made repeated successful flights. He started by imitating birds with flapping wings. Then he dropped that idea and went to a kind of fixed-wing hang glider. He made many different kinds of gliders: monoplanes, biplanes, and airframes of every conceivable shape.

In six years' time, Lilienthal had made two thousand flights, and he was starting to think about powered flight. But then, one Sunday afternoon, a crosswind caught him fifty feet in the air. The glider sideslipped, crashed, and broke Lilienthal's back. According to legend, before he died, he murmured, "Sacrifices must be made." The trouble is, he had said that before. It was a typical Victorian sentiment, and Victorian sentiment almost certainly tied the remark to his death.

In 1900 Wilbur Wright wrote a letter to the next great glider pioneer, Octave Chanute, asking for advice. In the oddest way, his language evoked both Lilienthal's death and Orville's illness four years earlier. Wilbur wrote:

> I have been afflicted with the belief that flight is possible.... My disease has increased in severity and I feel that it will soon cost me...increased money...if not my life.

Well, it was disease and not his belief in flight that eventually killed Wilbur. But the most elusive quest in the world is the search for the origin of an idea. The Wrights com-

Otto Lilienthal in flight.

bined invention with remarkably thorough study and laboratory work. They worked methodically and inexorably for years, and their powered flight in 1903 rested upon that labor. But it also went back to summer's end in 1896—to a time when Lilienthal died and Orville lived, to a time when two brothers became certain of what they were destined to do.[10]

Victorian sentiment colored descriptions of invention in the late nineteenth and early twentieth centuries. We still find its echoes in accounts of the invention of the vacuum tube. Perhaps by making theater of invention we actually provide an avenue into invention. The story of the vacuum tube begins with the *Edison effect*. That was the name given to a phenomenon that Edison observed in 1875, although it had been reported two years earlier in England. Edison refined the idea in 1883, while he was trying to improve his new incandescent lamp.

The effect was this: In a vacuum, electrons flow from a heated element, such as an incandescent lamp filament, to a cooler metal plate. Edison saw no special value in it, but he patented it anyway; he patented anything that might ever be of value. Today we call the effect by the more descriptive term *thermionic emission*.

The magic of the effect is that electrons flow only from the hot element to the cool plate, but never the other way. Put a hot and a cold plate in a vacuum and you have an electrical check valve, just like check valves in water systems. Today we call a device that lets electricity flow only one way a *diode*. It was 1904 before anyone put that check valve action to use. Then the application had nothing to do with lightbulbs.

Radio was in its infancy, and the British physicist John Ambrose Fleming was working for the Wireless Telegraphy Company. He faced the problem of converting a weak alternating current into a direct current that could actuate a meter or a telephone receiver. Fortunately, Fleming had previously consulted for the Edison & Swan Electric Light Company of London. The connection suddenly clicked in his mind, and he later wrote, "To my delight...I found that we had, in this peculiar kind of electric lamp, a solution."

Fleming realized that an Edison-effect lamp could convert alternating current to direct current because it let the electricity flow only one way. Fleming, in other words, invented the first vacuum tube. By now, vacuum tubes have largely been replaced with solid-state transistors, but they have not vanished entirely. They still survive in such modified forms as television picture tubes and X-ray sources.

Fleming lived to the age of ninety-five. He died just as World War II was ending, and he remained an old-school conservative. Born before Darwin published his *On the Origin of Species*, he was an antievolutionist to the end. Yet even his objection to Darwinism had its own creative turn. "The use of the word evolution to describe an automatic process is entirely unjustified," he wrote, turning the issue from science to semantics.

In an odd way, semantics marked Fleming's invention as well. He always used the term *valve* for his vacuum tube. In that he reminds us that true inventors take ideas out of context and fit them into new contexts. Fleming stirred so much together to give us the vacuum tube—not just lightbulbs and radio, but water supply systems as well.[11]

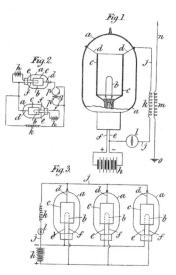

The theater in Fleming's discovery lies in his own account of it. Yet what he dramatizes is a process we have already talked about. The creative inventor takes ideas out of their original contexts and uses them in new ones. He turns bread mold into penicillin, coal into electricity, or, I suppose, lead into gold, because he is not constrained to keep each thought in its own container.

J.A. Fleming's 1904 patent drawing.

Another inventor, Thomas Sopwith, offers little theater. He was solid and workmanlike all his long life. The drama of Sopwith's life is cumulative, apparent only when we look back over our shoulder at what he accomplished. But let us go back to the beginning for his story. The airplane that finally brought down the "Red Baron," von Richthofen, was an English biplane called the Sopwith Camel.

Camels were maneuverable little airplanes, and my father, who flew them in France, told me they were tricky to fly. But with a good pilot, they were deadly in combat. In 1916 the Germans controlled the air over the Western Front. The Sopwith Camels challenged their dominance when they arrived in France in 1917. They were given the name *Camel* because the contour of the fuselage included a humplike cowl over the guns, in front of the pilot. And Thomas Octave Murdoch Sopwith manufactured them.

Sopwith was fifteen when the Wright brothers flew. He learned to fly in 1910, when he was twenty-two. By then, he had raced automobiles and speedboats, and he had done daredevil ballooning. In no time, he won flying prizes, and he used the prize money to start making airplanes. He was twenty-four when World War I was brewing. His first planes were used early in the war, and when the Sopwith Camel gave the air back to the Allies in July 1917, Sopwith was still under thirty.

He stayed with airplane manufacturing after the war. In 1935 he was made chairman of the Hawker-Siddley Group, and there he did a most remarkable thing. In 1936 he decided to produce a thousand Hawker Hurricanes on his own, without a government contract. War was brewing again, and if the British government was not ready, at least Sopwith was. Without his Hawker Hurricanes, England would have been laid bare by the Nazi bombers during the Battle of Britain.

But that was far from the last of Sopwith. After World War II he was involved in developing the Hawker Harrier—the first jet airplane that could take off and land vertically. The Harrier was quite prominent during the Falklands War.

Sopwith celebrated his hundredth birthday on January 18, 1988. The RAF sent flights of Sopwith's airplanes over his home near London: an array of airplanes from early flying machines to modern jets! It was a parade that represented the whole history of powered flight during the life of this remarkable man, who had an uncanny ability to read the future. He died a year later, at the age of 101.[12]

So where do inventive ideas come from? Sopwith's life no more answers that question than Gibbs' life does. However, it is worth looking at two stories that include intimations of how inventors found their way to invention. We shall begin the first with the Sperry-Rand Corporation trying to sue Honeywell in 1967.

Honeywell was making digital computers, and Sperry claimed that Honeywell owed them a royalty. After World War II, Sperry had bought the patent rights to ENIAC, the first digital electronic computer. Honeywell countersued Sperry. They made the extraordinary claim that Sperry's patent was invalid—that the digital computer had been invented before ENIAC.

Honeywell won its case six years later, and they did it by correcting history. People at Honeywell found their way back to the winter of 1937. John Atanasoff, a young physics instructor at Iowa State University, was struggling with the problem of mechanizing computation. Things were going badly this particular evening. Finally, in frustration, he jumped in

Looking Inside the Inventive Mind

his car and sped off into the night. Two hundred miles later, he pulled up at a roadhouse in Illinois to rest.

There it came to him. A machine could easily manipulate simple on-off electrical pulses. If computations were done in the either-or number base of two, instead of the conventional base ten, a machine could do calculations naturally. Sitting in that roadhouse, two hundred miles from home, he made the crucial step in inventing the digital computer.

Two years later, Atanasoff and a colleague named Berry started to build a computer. But in 1942 they were drafted, and the almost-complete computer was set aside without being patented. Meanwhile, the government started work on the ENIAC digital computer, which differed in some ways from Atanasoff and Berry's machine and in any case was bigger.

An unfinished, unpatented machine does not make a very strong claim in a priority dispute. But in this case there was a catch. One of the major inventors of ENIAC, John Mauchly, had known Atanasoff. Not only had they corresponded, but Mauchly had even visited Atanasoff in Iowa for a whole week in 1941. In the end, it was clear that the ideas that resulted in ENIAC actually were Atanasoff's.

Atanasoff did all his work with only six thousand dollars in grant money. But the military funded the ENIAC project. They were interested in making artillery firing tables, and they put a half million dollars into ENIAC. That was a huge sum in 1942.

So the next time you turn on your PC, the next time you spend thirty seconds doing some task that would have taken your mother or father all afternoon, think about a man clearing his mind one winter night in 1937. Think about a man gazing at a yellow line for five hours, until—suddenly—he was able to see through the dark.

The other story has to do with the problem of firing a gun—a cannon—from the deck of a rolling ship. In the War of 1812 warships fired banks of guns, broadside, at a nearby enemy. Seventy-five years later, ships carried long-range guns that fired from turrets. Aiming broadsides from rolling ships had been hard enough, but hitting a target a mile away was a whole different order of difficulty. In a typical test made in 1899, five American ships took turns firing at a ship hulk a mile away. After twenty-five minutes they had scored only two hits. By this time, ships' guns had telescopic sights and mechanisms that let gunners move them fairly quickly. The standard drill was to set corrections for the range, aim the gun roughly, then wait until the ship's roll put the target in the sights.

In 1898 an English admiral, Percy Scott, watched his men at target practice. All but one was doing miserably. That one gunner had evolved his own aiming tactic. He kept his eye on the sight and he moved the gun continuously until he could feel the synchronization between his aim and the motion of the ship. What he was doing was subtle, yet it took advantage of skills that most people already had. It coupled the man and the machine.

Scott adopted this technique and quickly set remarkable records in marksmanship. In 1900 an American naval officer, William Sims, met Scott in the Far East and learned all about his new technique. By 1905 the continuous-aim firing method had become standard U.S. Navy practice, but not before Sims had learned a grim lesson about innovation in organizations.

Sims' attempts to interest the U.S. Navy in continuous aiming initially met a brick wall. English equipment, he was told, was no better than ours. If the men could not hit the target, it was because their officers did not know how to train them. And they told Sims flatly that continuous-aim firing was not possible. Sims finally went straight to Teddy Roosevelt, who recognized something in him and decided to give him a try. He abruptly made Sims the chief inspector of target practice. By 1905 a single gunner scored more hits in a one-minute test than the five ships had previously done in twenty-five minutes.[13]

That unknown English sailor thought not about mastering standard technique but about how to do the job. Scott recognized the importance of what he had done. And Sims championed the idea.

Today, we ask ourselves how to shorten those three steps toward putting a good idea into play. How do we escape the mental straitjackets that keep us from seeing new possibilities? How do we give organizations the capacity to recognize value in invention? And how do we show people what that creativity can do for them? Not one of those tasks is easy. And we are back at our opening premise—that invention is a willingness to be surprised. For all three steps require that we open ourselves up to surprise, and that is surely one of the hardest things to do.

4

The Common Place

A contradiction swirls around invention. While invention flows from an uncommon quarter of the mind, it ultimately comes to rest in the day by day world where we live our lives. Invention defines the commonplace world that we all share. The creative imperative is a unique and wonderful thing, yet it grows in the common clay of coping and of play, and that is also where it comes to rest.

We celebrate the magnificent steam engines, airplanes, and cathedrals. But look around your room for a moment. When I do that I see paper, windowpanes, wood screws, a pencil sharpener, paint, and carpeting. Everything but the cat sleeping on the window ledge came into being after long sequences of invention by many people. Even the cat's subtle gestures and communications may be partly the stuff of my own contrivance.

When we look with the eye of the mind at the everyday world around us, we see how much human imagination has run riot through it. We realize how imagination has invested the basest elements of our lives with possibilities. Try counting the cost of the ordinary world in the coin of human ingenuity. Cartographers who invented the globe on my bookshelf gave me a way to visit Fiji, Chad, and Tibet—places where fortune is unlikely to take me. The simple crank mechanism on my pencil sharpener represents a huge leap of the mind that took place only about twelve hundred years ago (a matter we talk about in Chapter 11). Imagination has enriched every corner of those common places where we all live out our lives.

For example, a hassled secretary hacks out a living on a new IBM typewriter in 1951. The typewriter's ribbon ink leaves nasty smudges when she erases an error. In a burst of creative frustration, she goes home and invents a liquid for painting out mistakes. Its base is white tempera paint, the woman's name is Betsy Nesmith, and the liquid is an immediate hit with other typists. By 1956 she is running a cottage industry, mixing the brew for other secretaries. She calls the stuff Mistake Out, and her small son Michael (later to become one of the original Monkees) helps her fill hundreds of bottles a month.

By the time the number stretches into thousands per month she renames the product Liquid Paper. Then one day her mind wanders at work and she types The Liquid Paper Co. on a company letter. Her boss fires her, but no matter. The Liquid Paper Co. is putting out twenty-five million bottles a year when she retires as Chairman of the Board in 1975.

Another invention helps us see just how imagination interacts with the common place. To find the leap of the mind represented by the bicycle, we recall how psychologist Jean Piaget questioned young children struggling to understand it. Four year olds see the bicycle as a whole—a thing that goes entirely of itself. We listen as Piaget talks with a child:

How does this bicycle go?
    With wheels.
And how do the wheels turn?
    The bicycle makes them turn.
How?
    With the handle-bar.
How does the handle-bar make the bicycle go?
    With the wheels.
And what makes the wheels turn?
    The bicycle.

The child calls parts into the explanation and drops them just as quickly—for the bike is entire unto itself. When children reach six, they start referring to parts, but not in any orderly cause-and-effect way. You hear things like:

What makes the chain turn?
    The wheels.
What makes the wheels turn?
    Those brake wires.[2]

The six-year old is just starting to forge understanding by breaking the bicycle down into its component pieces. Most children are nearly nine before they can sketch a bike from memory and explain how the parts work. Piaget's children and their bicycles reflect the same circuitous process by which the inventor first created the bicycle. Piaget demonstrates how we replicate the inventor's original thinking as we learn to analytically decompose things into cause-and-effect sequences.

As inventors, we must see more than decomposed parts. We must also be able to put the parts back into a whole. We must be able to find our way back to the thinking that once marked our childhood, but, once we have, the truest answer to the question of how a bicycle works is that we and the machine become a single thing. Invention builds a bridge between analytical decomposition and the unity of the finished thing. The inventor must, at last, forget the wheels, chain, and sprocket of this wonderful device and, finally, sail down the road—riding the wind.

So imagination makes the common sand in our lives into pearls by rekindling a childlike ability to see the world in a common grain of sand. Look at the things around you that you might never have regarded as having been invented. Some are so elemental that you would think they had been around before the world began. Look in your purse or your wallet: Did someone actually invent the idea of replacing real goods and services with a medium of exchange? Are we right to view the abstraction of goods and services that we call *money* as an invention?

We certainly are, but the invention took place in several stages. The first stage dates to the dawn of recorded history. As people from different places bartered and traded their produce, they needed frames of reference to set the value of goods. So they started referring the value of everything to such everyday trade items as cattle or metal.

This process was generalized as early as 4,000 B.C. by using the weights of certain metals—usually gold, silver, or copper—as a reference. When the Bible uses the word *talent*, it is actually referring to a unit of weight. Fifty- or sixty-pound copper talents were a regular medium of exchange around the Mediterranean by the ninth century B.C.

The next stage was for governments to certify a particular medium of exchange. The Chinese did so in the eighth century B.C. when they put government inscriptions on certain exchange goods. The Ionians began inscribing talents a century later. Lydian Greeks minted the first coins just after 650 B.C. A coin takes the abstraction one step further by inscribing a valuable piece of metal with a guarantee that it has value, even though it is separate from actual goods. The phrase "rich as Croesus" refers to the fifth-century B.C. Lydian king Croesus, famous for minting coins. Greek coins soon reached an artistic level that would not be matched until modern times.

The next stage of abstraction was to replace valuable metal with paper money. When that happened, money became only symbolic of material goods with no value of its own. The Chinese government issued paper money in the twelfth century, but Europe has only dared to do so during the last three hundred years.

Now we have reached yet another stage in the process of abstraction. We call the whole process of barter into computers, and we reduce our balance of cows, autos, and labor to electronic record keeping. Our goods and services are converted into an agreed-upon figure of merit (call it the dollar or the yen) and weighed against one another in electric pulses. We still barter with each other, but the medium of exchange has, at last, become so abstract as to be completely incorporeal.

And that cannot be the end of abstraction. We have further to go. Subtler measures of value will emerge. The next medium of exchange will surely give trade value to immaterial information. The old term *intellectual property*, which tries to equate information and property, will give way to something more sensible. But what? It might even become clear in my lifetime; I certainly hope it will in the next generation.

Yet the common place of barter, trade, and real goods remains present in our use of money. That's why the physical  medium of money reflects our common aspirations with amazing directness. Turn coins about in your mind and you realize that they are the most durable written record of humankind. They reflect our myths and legends. They tell what we honor and what we find beautiful. American coins are much plainer and more static than most, yet even they display buffalo and Indians, wheat, liberty represented by a woman, bald eagles, and former presidents. They affirm our trust in God, variable though that conviction may be.

Coinage is an odd technology, since coins are three things at once. They are a historical record, they are often works of art, and, of course, they remain a claim to goods and services. Since (until recently) the metal used to make coins had value, that claim can linger long after their issue. Bronze coins minted by the first-century Roman Emperor Domitian were still being used in Spain as late as the seventeenth century. Philip IV finally called them in and had them restamped.

The first time I ever held an ancient coin in my hand, its expressiveness took my breath away. It was a late second-century B.C. Roman denarius. Jupiter's picture was on the front and, on the back, was a quadriga, a Roman war chariot. The rim was chipped all the way around, a result of the old way of making coins. After the silver blank had been stamped with the images, fragments were then nipped out of the unstamped rim until the coin had the right weight. That finicky little silver-saving process gave us the word *penny-pinching*. It also showed that the coin was solid, not just silver-plated.

A century later, the new gods, the imperial Caesars, replaced the old household deities on the front of the denarius. The republic was now an empire. (The old war chariot survived for a while; Rome kept its interest in war.)[3]

I study a handful of contemporary Cayman Island coins and wonder what Caymanians think about. The coins tell me that they voluntarily claim membership in the British Commonwealth. Queen Elizabeth graces all their coins. But on the other side are boats, birds, shrimp, and the Caymanian national symbol, the turtle. Anyone looking at these coins a thousand years from now will be able to see the islands and their natural beauty through Caymanian eyes.

Money, after all, represents the work of our hands—our technology. Our interest in money has a component that is a great deal more honorable than mere greed. Money represents what we do. And what we do is what we are. A curious biblical remark says that our heart will be where our treasure is. It sounds cynical at first, but it makes perfect sense when we see money as a kind of condensed representation of ourselves. In the end, it is not surprising that we reveal our hearts in this most peculiar art form. We say who we are and what we value when we coin money.

At about the same time that ancient thinkers conceived of the talent as a medium of exchange, artisans in that same eastern Mediterranean world were also beginning to create another new technology of the common place. The first glass products appeared about forty-five hundred years ago.

The evolution of both technologies has been glacial. Like coins, decorative glass and glass tableware (bowls, goblets, dishes) have been recurring forms of high art since the Hellenistic era. Yet something as wonderfully utilitarian as a uniform windowpane was not generally available until after paper money was.

The ancient Egyptians and Greeks made crude glass decorations, but today's basic soda-lime glass—made of sand, limestone, and sodium carbonate—is only about two thousand years old. The first glass of any real quality was made in Hellenistic North Africa around 300 B.C. Soda-lime glass followed quickly on its heels, and both the Hellenistic Greeks and the Romans after them made it into fine tableware. Such tableware remained the most common glass product for a long time.

The stained-glass art of the Gothic cathedral was so sophisticated that we might think glass handling had reached a high level of perfection by that point. Actually, the skill that had been perfected, one that seems to be beyond anything we know how to match today, was *coloring* the glass. A medieval window admitted light, but it was seldom smooth enough to provide a clear view. Cathedral windows were meant to admit light, but they did not even try to display the outside world. What they did do beautifully was to provide illuminated Bible stories for the faithful, who for the most part could not read.

Typical early-nineteenth-century crown-glass-making operation, from the 1932 *Edinburgh Encyclopaedia*.

Medieval glassblowers made two kinds of flat glass sheets. By one technique, an artisan would blow a large cylinder, then split the cylinder open and flatten it out while it was still hot. The second flat glass sheet was called *crown glass* and was made by spinning molten glass and letting centrifugal forces spread it out from a central point. Crown glass was the most common form of flat glass for centuries. It underwent considerable refinement, but even as late as 1800 most domestic windows still displayed the characteristic umbilical imperfection, called a *crown*, at their centers.

The French had developed the superior, but expensive, plate glass process in the latter part of the eighteenth century. First, molten glass is poured out in a mold; then the glass needs expensive grinding and polishing. To provide you and me with cheap, high-quality domestic windowpanes, molten glass had to be rolled out in smooth, continuous sheets, and that couldn't be done until we had modern process machinery. It was in the early 1800s that the first inexpensive rolled window glass became available. That was less than two centuries ago.

The windowpane again reminds us how much inventive genius is embedded in taken-for-granted technology. It is the result of complex high-temperature chemical and mechanical processes. Yet few things give our daily lives the soul-settling grace of these unobtrusive bridges to the outer world. Like so much really good technology, this one is at its best when it is completely invisible.

About the same time medieval artisans were creating Bible textbooks of light and color in the walls of their cathedrals, the use of coal was just becoming commonplace in the European economy. Isolated reports told of coal burning by the ancients, yet Marco Polo had still been surprised by Chinese coal burning in the late thirteenth century. He would also have been surprised if he had traveled to northern Europe and England. By then, the English were using coal for smithing, brewing, dyeing, and smelting. They had even started exporting some to France.

By the thirteenth century, millwrights had spent two hundred years consuming the European forests to make their waterwheels and windmills. Wood was becoming too precious to use as a fuel. At first it was replaced by surface coal (often called sea coal because the more obvious outcroppings were found on the coast). By far the largest deposits of sea coal were in England. The reason we do not bring coals to Newcastle is that Newcastle, in particular, was surrounded by huge fields of sea coal. Coal was mined in open cuts, thirty feet deep, so Newcastle was soon girdled by a dangerous maze of water-filled trenches.

Sea coal was filthy stuff, loaded with bitumen and sulfur. It created environmental problems from the start. An anonymous fourteenth-century balladeer vented his anger at its use:

Swart smutted smiths, smattered with smoke,
Drive me to death with din of their dints,...
The crooked caitiffs cryen after col! col!
And blowen their bellows that all their brain bursteth.[4]

But the medieval population explosion drove people to use this foul fossil fuel anyway. For a hundred years medieval environmentalists fought with medieval industrialists over its use. Then famine and plague ended their argument until the middle of the fifteenth century. The repopulation of Europe drove people back to coal, but now they were armed with new techniques gained from mining metal.

It takes a hot, clean-burning flame to smelt metal. That was once done with charcoal—a fairly pure carbon residue left after the volatile components had been burned off, wood. So when wood shortages reappeared, they were magnified not only by rising populations but by the rising demand for metals as well. As miners followed coal seams down into the earth, mining it the way they had learned to mine metal, they finally reached the much cleaner hard coal we use today. And that deep-mined coal also served to replace wood in the smelting process.

Coal and metal drove miners deeper into the earth until in the seventeenth-century they were stopped by the water table. With their appetites whetted for fuel and metal, our forebears became desperate to feed their hunger. By now this is all too familiar. Human ingenuity creates new human appetites, which are eventually met by new ingenuity. It is as frightening as it is heartening to see how we find eleventh-hour means for keeping those appetites sated.

Of course the deforestation that brought about the need to take up coal was driven by far more than just the construction of water wheels and windmills. Northern Europe had been built upon its early abundance of wood. Much is written about the economics of wood, but we know all too little about the tools that made windmills, Chippendale chairs, or early American desks. The catalog of a Smithsonian Institution collection of hand woodworking tools, published several years ago, helped to remedy that.[5] These tools date from the seventeenth century up into the early twentieth century—a period spanning a little more than a century on either side of the Industrial Revolution—and they powerfully mirror their times.

Most woodworking tools survived this period without changes to their basic character. Seventeenth-century planes, saws, clamps, and chisels changed in their outward appearance, but it is easy to recognize them for what they are. Only one fundamentally new tool arose in that period, and it's one I never would have thought of: the gimlet auger, of all things—a tool we seldom see today.

By the nineteenth century the quality and extent of metalwork in wood-working tools had greatly increased. We find all sorts of fine screw fittings and adjustments that did not exist two hundred years earlier. Hand tools generally began to reflect a new style. It is almost as though the fine lines of the furniture they were making were rubbing off on them. The sinuous shape of today's axe handle—with the end curving to give the user a better hold—was in place by the nineteenth century. So was the comfortable pistol grip you expect to see on a carpenter's saw.

These changes began in the eighteenth century, when people began to see tools as instruments. Perhaps the breath of eighteenth-century science was blowing into the cabinetmaker's shop. During these years, a profusion of stylish new precision measuring instruments arose to support hand tools, such as fancy calipers and dividers. But the tools themselves also began to look like scientific instruments. We see a hundred-year-old drill brace made of heavy brass with clutches, ratchets, and shiny japanned handles. Do you remember the lovely old wood plane, made of brass and tool steel—the array of adjustment knobs, the beautifully formed wooden handles?

We still do woodwork. We still seek the essential pleasure of shaping wood with our own hands, but now we use power tools. Tools are intimate possessions. They reflect the way we see ourselves. I may be nostalgic for turn-of-the-century tools and for the lost harmony of wood, brass, steel, and form, but when the chips are down, I am a creature of this world. When I want to make a hole, I reach for a power drill.

Consider next a truly elemental technology of the common place. Like wood, textiles are the subject of a great deal of economic history. A nearly invisible item was, for centuries, essential in the use of textiles: the ordinary straight pin. An old nursery rhyme says, "Needles and pins, needles and pins, when a man marries his trouble begins." The lowly dressmaker's pin was once a metaphor for commonplace household necessity. Most people made their clothes at home in the early nineteenth century, and dressmakers absolutely needed pins. But pins were hard to make. They were once made by hand in home production lines, with each person doing one operation. In his poem "Enigma," the eighteenth-century poet William Cowper described a seven-man pin-production line:

> One fuses metal o'er the fire;
> A second draws it into wire;
> The shears another plies,
> Who clips in lengths the brazen thread

For him who, chafing every shred,
Gives all an equal size.

A fifth prepares, exact and round
The knob with which it must be crowned;
His follower makes it fast;
And with his mallet and his file
To shape the point employs awhile
The seventh and the last.

But pin making was actually more complex than Cowper's seven-stage process suggests. Eighteenth-century economist Adam Smith described *eighteen* separate steps in the production of a pin. Small wonder, then, that pin making was one of the first industries to which the early-nineteenth-century idea of mass production was directed.

The first three patents for automatic pin-making machines were filed in 1814, 1824, and 1832. The last of these, and the first really successful one, was granted to an American physician named John Howe. Howe's machine was fully operational by 1841, and it has justly been called a marvel of mechanical ingenuity. It took in wire, moved it through many different processes, and spit out pins. It was a completely automated machine driven by a dazzlingly complex array of gears and cams.

When Howe went into production, the most vexing part of his operation was not making pins but packaging them. You may have heard the old song that goes: "I'll buy you a paper of pins, and that's the way our love begins." Finished pins had to be pushed through ridges in paper holders, so that both the heads and points would be visible to buyers. It took Howe a long time to mechanize this part of his operation. Until he did, the pins were sent out to pin packers, who operated a slow-moving cottage industry, quite beyond Howe's control.[6]

So we glory in the grander inventions—in steam engines and space-ships, gene splicing and fiber optics. The pin is one more technology that serves us by easing the nagging commonplace needs that complicate our lives. If the TV and the door handles both disappeared from your house, which would you replace first? Making the lowly dress-maker's pin easily available was a substantial blessing to nineteenth-century life and well-being.

We also lose track of the commonplace dimension when we talk about the technologies of getting from place to place. After we've dealt with steamboats, automobiles, and airplanes, we discover that we've overlooked the transportation medium that has touched every one of our lives. We mentioned the bicycle at the outset, as it represents commonality of thinking. Now let us look at as it represents commonality in transportation.

The history of the bicycle is curiously tied to that of the horseless carriage. Together they represent ways that the poor and the wealthy achieved freedom of movement. The early nineteenth century saw all kinds of new steam-powered vehicles. At first steam carriages competed with locomotives, but the railways won the battle because they made transportation inexpensive in a way steam carriages could not.

Trains were confining, and people wanted freedom to travel the roads as they pleased. The new dream of rapid movement had to be individualized. If the answer was not the steam carriage, then maybe it could be the bicycle. Between 1816 and 1818 Scottish, German, and French makers came out with primitive bicycles. They all seated a rider between a front and a back wheel with his feet touching the ground so he could propel himself with a walking motion. This form of bicycle was not new. Such bikes are found in Renaissance stained glass, Pompeian frescoes—even in Egyptian and Babylonian bas-reliefs.

But around 1839 the Scottish maker Macmillan added a feature to his "hobbyhorse," as he called it. He added a pedal-operated crank to drive the back wheel—like the pedal-operated chain drive on your bike. Oddly enough, the idea didn't catch on. Not until 1866 did pedals appear on the *front* wheel, like on the tricycles we rode as children.

Bicycle use took off after the invention of the front-wheel pedal. But it also led to larger and larger front wheels. The bigger the wheel, the further the bike would move on each turn of the pedal. That led to the dangerously unstable bicycle depicted in so many Currier and Ives prints—the one with the huge front wheel and the tiny back one. In its developed form it was called the *ordinary bicycle,* but it was nicknamed the *penny-farthing* because its wheels looked like large and small coins.

The ordinary was so tricky it finally gave way to the so-called *safety bicycle*—the modern bike with two equal wheels, the back one driven by a chain and sprocket. The safety bike resembled MacMillan's hobbyhorse design forty-six years earlier. It went into production around 1885. It soon replaced the ordinary and has been the basic bike design ever since.

So the modern bike entered the twentieth century along with the new gasoline automobiles. It freed those people who could not afford

The transition to the new safety bicycle is dramatically illustrated in this heading for an article in the September 1896 *Century Magazine*.

cars. Now they too could go where they pleased. And oh, the sense of freedom we felt as children when we got our first bikes! They let us fly like the wind and go where we wanted. They were wonderful things.

The technologies of the common place remain invisible only until we are separated from them. The literature of exploration and settlement reveals all the ingenious means people find for re-creating the most basic of our technologies. Ellen Murry, then at the Star of the Republic Museum at Washington-on-the-Brazos, Texas, has written about the early technologies of this rough land.

The Republic of Texas seceded from Mexico and became a separate nation in 1836. It was a wild, unsettled nation. Early Texans were intimate with untimely death. Mourning and memorializing death were basic social activities. A disturbing level of attention was paid to the crafts of preparing, displaying, transporting, and burying the dead. With death so commonplace, women sustained life by marrying in their mid- to latter teens and by raising lots of children. Normally, six or seven kids survived after murderous infant mortality took its toll. Texas frontier women—often managing despite their husbands' long absences—did the child rearing, educating, and civilizing.

These settlers had little access to any developed medical technology. They fought illness by trying to rid the body of whatever ailed it. They embraced the medieval idea of curing by bloodletting, emetics, and laxatives. "Puke and purge" was a saying that began and ended most medical treatment on the Texas frontier.

People did recognize that unsifted whole-wheat flour was good for the digestion. A major apostle of that notion was Sylvester Graham—promoter of the Graham cracker. He also suggested that it reduced alcoholism and damped the bothersome sex drive. Bathing, too, was a form of medical treatment. It had little other place in everyday life. In 1840 a writer denounced the bathtub as "an epicurean innovation from England, designed to corrupt the democratic simplicity of the Republic." Early Texans washed their hands and faces before meals, but it was normal to go a year or more between baths.

The Common Place

Tobacco, especially chewing tobacco, was an early Texas fixation. Children were taught to use the stuff. Cuspidors were universal furnishings. A visitor to the Texas Congress said, "The way the members were chewing Tobacco and squirting was a sin to see." And an Austin church posted this notice:

Ye chewers of the noxious weed
    Which grows in earth's most cursed sod,
Be pleased to clean your filthy mouths
    Outside the House of God.

The Republic of Texas lasted less than a decade. Any way you hold them up to the light, the people who formed it were tough, independent, adaptive, and idiosyncratic. We get to know them when we look at their daily means—the rough-hewn technologies by which they carved freedom and the good life out of a harsh, and seemingly infinite, land.[7]

And finally, a brief look at one of the most common technologies of them all, the technology of waste disposal. Every time I teach the history of technology some student tells me, often with a salacious grin, that the flush toilet was invented by a nineteenth-century Englishman named Thomas Crapper. Well, he did not really invent the flush toilet, but his name does indeed hover over its history.

The modern flush toilet consists of three essential components: a valve in the bottom of the water closet, a wash-down system, and a feedback controller to meter the next supply of wash-down water. The first two were incorporated in the toilet developed by a courtier of Queen Elizabeth, a poet named John Harington. The third element, the float-valve feedback refill device, was added in the mid-eighteenth century. The flush toilet was, in fact, an important landmark in the Industrial Revolution. It was closely tied to the new technology of steam-power generation, since the important concept of automatic control of liquid levels arose both in steam boilers and in the tanks of these new water closets.

Thomas Crapper was a real person. He was born in Yorkshire, England, in 1837, long after the first flush toilets came into use. His wonderfully tongue-in-cheek biography is told by Wallace Reyburn in a book entitled *Flushed with Pride*. Thomas Crapper apprenticed as a plumber when he was still a child. By the time he was thirty he had set up his own business in London. He developed and manufactured sanitary facilities of all sorts until his death in 1910. He held many patents and was, in fact, an important and inventive figure in the creation of modern water closet systems.

But did he really give his name to these systems? Reyburn claims that many American soldiers in World War I were off the farm. They had never seen anything like the classy English water closets and called them by their brand name, much as the English call a vacuum cleaner by the brand name Hoover.

The problem with this explanation is that the word almost certainly derives from the thirteenth-century Anglo-Saxon word, *crappe*. It means "chaff" or any other waste material. The modern form of the word was certainly in use during Thomas Crapper's life. So he was not the inventor of the flush toilet, and it is unlikely that he really gave it his name. What he did do was to carry the technology forward.[8]

The Thomas Crapper story points out something historians have to guard against. Now and then a really good story comes along—one so well contrived that it should be true, even if it is not. Who wants to admit that no apple ever fell on Isaac Newton's head or that George Washington did not really chop down the cherry tree? What humorless pedant wants to insist that Thomas Crapper did not really invent the flush toilet?

Since waste removal is a universal common denominator, anthropologists have used it to answer a question any historian inevitably asks: What was the texture of life in a given period? What was it like to live in ancient Rome, in a medieval castle, or, in this case, in an early Texas mansion? Life was primitive inland, but Galveston was another matter. Galveston was Texas's major port and a town marked by some grace and elegance. Yet we gain one of our best views of the lives of Galveston's wealthy by diving into a old privy.

Ashton Villa is a mansion that survived the terrible 1900 Galveston hurricane. The house was built by James Brown, a wealthy businessman who kept slaves before the Civil War. It is very well made because he used the work of slave craftsmen instead of manufactured matériel. Urban archaeologists Texas Anderson and Roger Moore have shown how to wring the mansion's story from it. The old privy, long since covered over and forgotten, becomes their window into the past. More than an outhouse, it was also a general trash dump. A huge hole in the ground, no longer septic, it contains layers of trash that reveal the quality of life from the late 1850s up through the late Victorian period.

Galveston rode out the Civil War better than most of the South, and so did Brown. After the war he furnished the house with fine European porcelain. His family ate inch-thick T-bone and porterhouse steaks; they disdained chicken and pork. They drank elegant wines and cognacs, but not hard liquor. The ladies imported French perfume and expensive facial astringents. Brown's business involved selling new technology to the American West. His mansion displayed all the latest stuff—the first flush toilets in Galveston, and the first electric lights only a few years after Edison introduced them.

Sifting through century-old detritus, we begin to sense the finery and feel of the place and to know the actual people. We begin to understand the combined tyranny and vision that Brown represented. To say merely that he exploited slaves or that he brought technology to the West is like trying to know baseball from sports-page statistics. But the intimacy of an accurate look into the drawing room or the servants' quarters is an understanding of a whole different order when we have the wits to look at them through a trash heap.

Items found in the Ashton Villa privy pit included fine china, perfume, wine and medicine bottles, and bones from the best cuts of meat. (Photos courtesy of Texas Anderson)

So history is revealed where history is made. When we look for history we ultimately find it by looking in the commonest of common places. Kings and emperors only appear to shape our world. Technology, culture, and the ghosts of history itself are born at our human center. To know them, we must know the common place.

> *If we had a keen vision of all that is ordinary in human life, it would be like hearing the grass grow or the squirrel's heart beat, and we should die of that roar which is the other side of silence.*
> —George Eliot,
> *Middlemarch,*
> Chapter XX, 1871

5

Science Marries
into the Family

The old Latin word *scientia* was not much used to designate ordered knowledge until recent times. Galileo would not have called himself a scientist, nor would Newton or Leibniz. Even two hundred years ago Lavoisier still called himself a natural philosopher. Yet each of those people contributed to the radical change that turned natural philosophers into today's scientists.

The change was complex. Roughly speaking, it could be called the evolution of the scientific method. It began during the 1480s, when the new printed books first included accurate illustrations of an observed world. Until then, first Christians and then Moslems had adhered to the Platonist idea that truth is to be reached by deduction far more than by observation.

Throughout the sixteenth century the new observational sciences of botany, anatomy, descriptive geometry, geography, and ethnography all took form. In a series of bold steps the new media of print and illustration wrenched a world still shaped around Platonist thinking. Then, in the early sixteen hundreds, Galileo Galilei and Francis Bacon in particular codified the changes that had been afoot. Galileo did more than anyone to establish the methods of the new science, and Francis Bacon framed its philosophical stance.

In 1620 Bacon wrote down the new view of nature in unmistakable terms in his *Novum Organum*. He directly contradicted the Platonists' belief that truth is to be found in the human mind when he said, "That which nature may produce or bear should be discovered, not imagined or invented." For over a century, a new breed of scientists had been learning how to take better account of empirical fact than medieval scientists had done. Now Bacon told us flatly that this was the only proper way to practice science.

After 1600 Europe gained two new tools of inquiry, both of which led away from medieval thinking. The shift to observational science was certainly strengthened by new kinds of measuring instruments. Clock making was a technology that had led to a new precision in mechanisms. The seventeenth century gave us the telescope, the thermometer, the vacuum pump, and the microscope.

But a second major force was also afoot, and its relation to the shift in scientific method was more complex. New forms of practical mathematics offered lay people means to perform calculations. Mathematics is an inherently Platonist endeavor. It is done within the human mind, with minimal recourse to the outside world. At the same time it is also the ultimate means for speaking absolutely objectively. Mathematical outcomes admit very little meddling by our subjective whims.

Armed with new mathematics and new instruments in the laboratory, science was not only in a better position to deal with external realities. It was also in a position to take part in the work of technology (which is, after all, the means for *manipulating* external realities). Natural philosophers of an earlier age had seen no relation between their work and the business of making things. Now a wedding would unite the activity of describing the world and that of shaping the human environment.

One of the early progeny of this marriage was architecture. Medieval masons had taken stonecutting very far from the one-dimensional stack of rock that made a Greek column, or the two-dimensional shape of a Roman arch. They had learned to shape masonry into complex rib work on the roofs of vaults, into helical stairways, and into arches that intersected at strange angles. Yet they had done this work without using formal geometry.

Geometry had been central to medieval scholasticism, but it had been an exercise in logic, not a means for making things. Masonry was crying out for geometric assistance, but the inevitable joining of mathematics to masonry did not come about until the end of the fifteenth century. When it did, a new baroque architecture emerged—one based on exact geometrical methods. Then architects started using precise intellectual apparatus to design magically spatial forms: barrel vaults, biased arches, helicoids, and embellished versions of the medieval trumpet squinch. A trumpet squinch is a conical arch that emerges smoothly where two walls meet in a corner. It is a beautiful complicated support for the floor above. Just imagine trying to cut the stone blocks that can be piled into such a form![1]

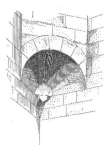

A trumpet squinch, as pictured in the *Dictionnaire Raisonné de l'Architecture*, 1868.

The wedding of modern, math-driven, observational science to fluid baroque architectural forms was only one of many such marriages. Those unions bore remarkable fruit, but make no mistake: They were weddings of opposites, reminiscent of the gladiatorial combat between a competitor using a net-and-trident and one carrying a shield and short sword—a perfect balance of evenly matched, but quite different, adversaries. The partners retained their own identity, and they sustained a level of running combat.

Throughout the last four centuries, science and technology have provided the essential tension needed to drive the mind. Today's propagandists for schools and industries like to celebrate the cooperation of these two oddly matched contestants. But what science and technology have achieved has sprung from something far more deep-seated than mere cooperation. Our complex modern world was created by these slightly irritable old companions who know, but do not like to admit, that they can no longer live without each other.

Galileo, who lived from 1564 to 1642, embodied all these contradictions. When he was young, his English contemporary Robert Burton used these words to describe Aristotle's idea of how objects fall (versification is mine):

> There is a natural place for everything to seek, as:
> Heavy things go downward
> Fire upward,
> And rivers to the sea.[2]

It was in the nature of falling, Aristotle had insisted, that heavy objects seek their natural place faster than light ones—that heavy objects fall faster.

Galileo took an interest in rates of fall when he was about twenty-six years old and teaching mathematics at the University of Pisa. It seemed to him that if Aristotle was right, a body unimpeded by air resistance should fall at a speed proportional to its density. He decided to test his elaborated Aristotelian theory with an experiment.

But there was no tradition of either making or describing controlled scientific experiments in Galileo's day. So Galileo's description of his experiments was quite skimpy by today's standards. Indeed, he doesn't actually claim to have *done* the experiment at all. Throughout his dialogs in *Two New Sciences* he peppers the pages with statements that hover between the language of deduction and that of observation.

> Surely a gold ball at the end of a fall through a hundred braccia will not have outrun one copper by four inches. This seen, I say, I came to the opinion that if one were to remove entirely the resistance of the medium [in this case, the air] all materials would descend with equal speed.[3]

Reading his dialogs, you never quite know if you are reading about Aristotelian observations or Platonist thought experiments. He seems to have dropped two balls, one of oak and one of lead, from a tower. But what sizes and what tower? If it was a real experiment from a real tower, we can be pretty sure it was the Leaning Tower of Pisa. Since Galileo spoke the language of Platonist discourse, he left historians of science wondering whether he actually *did* the experiment. Maybe he just reported what *should* have happened. Certainly that is how the Platonist philosophers around him studied nature.

One result of the experiment surprised Galileo and one surprises us. Galileo found that the heavy ball hit the ground first, but by only a little bit. Except for a small difference, which he correctly attributed to air resistance, both balls reached the same speed. That surprised him; it forced him to abandon Aristotelian writings about motion. If he really did the experiment, it was a turning point in the practice of science.

But what surprises *us* is what Galileo said happened just after he released the two balls. The lighter ball, he said, started out a little bit faster than the heavy ball. Then the heavy ball caught up. That sounds crazy in the light of known physics. So physicists Thomas Settle and Donald Miklich reran the experiment in front of a slow-motion movie camera. An assistant held two four-inch-diameter iron and wooden balls at arm's length, as Galileo would have held them to clear the wide balustrade at the top of the Pisa tower. A close study of the film proved that when someone tries to drop both balls at once, their strained muscles fool them. They consistently let go of the lighter one first. So what Galileo accurately reported is what really would have happened, and we are left with no doubt that he actually did the experiment.[4]

Galileo's tower experiment turns out to have been no fable after all. It was a very early controlled scientific experiment—the means Galileo used to become the first real challenger of Aristotle. The meaning of "Aristotelian science" can become confusing because Platonist medieval science had been built upon the body of knowledge that Aristotle had provided, much of it correct and some of it in error. But Medieval science then lost sight of the observational methods that Aristotle had created to gain his knowledge. The irony is that Galileo used those Aristotelian methods to correct the canon of Aristotelian fact.

So a new breed of scientist began rewriting physics.

Take the strongly Platonist saying "Nature abhors a vacuum." That had been a major truism for the natural philosophers of Galileo's time. They held it up to the light asking what it said about the nature of things. Nature demonstrated her abhorrence clearly enough when, for example, one used a drinking straw. Nature tried to get rid of the vacuum by driving liquid up the straw.

A story is told about several Florentine engineers trying to draw water from a deep sump. Try as they would, they could not get the water to rise more than thirty-two feet. Today we know that atmospheric pressure can push water just that far and no further; but no one knew that yet. So they went to Galileo and asked what was going on. Galileo wryly replied that nature's abhorrence did not appear to extend beyond thirty-two feet.

What Galileo didn't tell them was that he was also struggling, even as they asked, to understand air pressure and vacuum. He hired a young assistant named Evangelista Torricelli in 1641, just three months before he died. Two years later, Torricelli invented the barometer and estimated atmospheric pressure. We honor him today by naming the torr, a unit of pressure, after him.

Otto von Guericke, an influential citizen of Magdeburg in Saxony, had also become interested in the atmosphere. He had studied the work of Galileo and Torricelli, but he was also involved in the administration of Magdeburg. In fact, he became the city's mayor in 1647. About this time he invented a vacuum pump and did some spectacular things with it. In 1654 he gave the citizens of Magdeburg a remarkable lesson in the force of the atmosphere. He made two hollow hemispheres, twenty inches in diameter, and fit them tightly together into a sphere. Then he pumped the air out of the sphere and let sixteen horses—eight on each side—try to pull them apart. The horses couldn't do it. It would have taken a force over two tons to separate the halves.[5]

That may have been more showmanship than science, but it served its purpose. Guericke had shown the world that seemingly insubstantial gases could exert astonishing forces—forces that promised to be harnessed one day. During the rest of the seventeenth century, all kinds of people struggled to find a way to make use of these forces. In 1698, Thomas Savery finally made a workable pump that used the vacuum created by condensing steam. Just a few years later Thomas Newcomen showed us how to make a steam engine on the same principle, and suddenly a new power-generation game was afoot.

Guericke's strange little sixteen-horse experiment was in a sense a genuine forebear of our modern power plants, which generate millions of horsepower. For now science was truly in a position to influence the course of technology.

Francis Bacon had not only called people to reject Platonist thinking but also officiated at the wedding service between science and technology when, in 1620, he issued this wonderfully provocative challenge: "Now the empire of man over things is founded on the arts and sciences alone for nature is only to be commanded by obeying her." That was radical advice in 1620 because natural philosophers and the people who built things lived in such different worlds. The outcome of the wedding of science and technology would not be clear for another two centuries and it would be a full generation before the first new technology was born as the direct result of a physical discovery. When that occurred, it was because people wanted to improve the accuracy of clocks.

Any mechanical clock depends on an *escapement*—an inertial mechanism that moves back and forth in a steady rhythm. When Galileo was still a young man, in 1585, the medieval verge-and-foliot type of escapement, illustrated in the figure at the top of the next page, was the one most commonly in use.[6]

That year, 1585, Galileo showed that the period of oscillation of a gently swinging pendulum was always the same, regardless of the amplitude of its swing. The pendulum stood to make an ideal escape-

Modern replica of a verge-and-foliot clock from Galileo's time. The foliot is the weighted crossbar at the top. It swings back and forth in a rhythm dictated by its inertia. The verge is the vertical rod with two small pallets, driven by the foliot. The pallets alternately engage and then release the pinned gear at the top, and it sets the pace of the clock's movement.

ment, because it always swings at the same speed, even while it runs down. It also conserves much of the energy of its movement during each stroke, while the energy of the foliot was completely lost on each stroke. Verge-and-foliot clocks ran down faster.[7]

The year before Galileo died, his son Vincenzo built the first clock that used a pendulum escapement. (Lest the name cause confusion, I should note that Vincenzo was also the name of Galileo's distinguished father.) Later in the seventeenth century the Dutch and English physicists Christiaan Huygens and Robert Hooke followed Vincenzo's work with improved theories of the pendulum and really good pendulum-regulated clock designs.

Since then, technologists have listened more and more carefully to Bacon's assertion that we must understand and obey nature if we are to command her. Today's engineers are trained to the teeth in science. Our aim is still to make things, but to do that we had better understand how to learn and submit to nature's laws before we try to control her.

More and more scientists began consciously following Bacon's advice during the seventeenth century. Robert Hooke, born in 1635, did more than anyone to demonstrate what Bacon had meant. Hooke was a generalist of astonishing range who had important and lasting things to say about optics, mechanics, geography, architecture, materials science, clock making, paleontology, and microbiology. He was a virtuoso scientist with one foot solidly planted in the technologies around him.

Isaac Newton was only seven years younger than Hooke, but he was far less clearly shaped by Bacon. Newton worked alone with severe, rigorous abstraction from the technologies, which he saw as worldly distractions. He tried to endow science with the purity of mathematics, and he valued intensity and rigor far more than he valued Hooke's breadth of understanding. The poet Alexander Pope must have understood, at least on some visceral level, what was happening when, in his Epitaph for Isaac Newton, he wrote about the specialized science Newton was creating: "One science only will one genius fit; / So vast is art, so narrow human wit."

Newton and Hooke eventually came to blows. When Newton turned to optics in 1675, he had little to say about the important work Hooke had done in the field—in particular, Hooke's development and use of the compound microscope. Hooke gently objected to the omission, and Newton counterattacked in black fury. Newton's anger far outreached any issue at hand. Hooke had been a prominent lifelong member of the Royal Society. Newton became the society's president only after Hooke died, in 1703. Then he set about reshaping it as well. Part of that reshaping was a systematic effort to bury Hooke. During Newton's twenty-four-year presidency, many of Hooke's papers were lost, his apparatus was allowed to rust away, and his name was not mentioned.[8]

Schematic sketch of a typical pendulum escapement mechanism.

Newton was the great intellect of the age, and his antipathy toward Hooke flowed from absolute conviction. Despite Newton's sanctimony about seeing further by standing on the shoulders of giants, he certainly regarded his immediate Baconian forebears as midgets. To his defense, it must be said that science cannot always be closely coupled to technology. At some point Hooke's Renaissance-man breadth would have to yield to more concentrated and specialized forms of science.

Newton actually did a great deal toward moving science back inside the human head. But Bacon's idea that science has to serve technology has strongly returned in the twentieth century—perhaps too strongly. Still, as it has done so, historians have finally begun rediscovering Robert Hooke's astonishing scientific scope and stature. Doing so has meant holding it up to the light of the residual argument between the Platonists and the Aristotelians that continued all through the eighteenth century.

The conflict of scientific means—Baconian versus alchemical, Platonist versus Aristotelian—was clearest in the study of chemistry, or rather the Platonist *alchemy,* which grew into chemistry. By Newton's time scientists were repudiating alchemy, but its hold on the Western imagination was still immense. It is significant that Newton himself remained an alchemist and wrote voluminously on the subject.

When we think of alchemy it is all too easy to imagine magicians trying to change lead into gold. Real alchemy was simply the study of chemistry as it had been undertaken from the third century B.C. all the way through the next two thousand years. The word probably comes from the Greek *kemiya,* which meant "to transmute or change matter." For that is what alchemy, no less than modern chemistry itself, has always been concerned with—changing substances into other substances.

Alchemy took its form after Aristotle adopted an idea that had been argued a century before him by Empedocles—that all matter combined the four elements of earth, air, fire, and water. Aristotle guessed that these elements could be changed (transmuted) by the action of heat and cold, or by dampness and dryness. Aristotle's ideas were developed first by the Greeks and then by Arab scientists. From time to time alchemy mired itself in metaphysical razzle-dazzle. The practical Romans had no taste for it at all. So as civilization spread north into Europe, alchemy all but vanished until the thirteenth and fourteenth centuries, after scholars had begun to reread the old Greek and Arabic texts.

Alchemy promised wealth to anyone who figured out how to transmute the baser metals into gold. It might seem a waste that so many alchemists devoted their lives to that, but the spin-off was enormous. By trying to understand transmutation they learned about practical metallurgy, about extracting metals from ores, and about chemical reaction. Their results were reported in terms alien to our ears, but the late medieval chemists were surprisingly able metallurgists.

Beginning in the sixteenth century, more accurate measurements began to provide alchemy with a specificity it had never previously had. By the late seventeenth century chemists now focused on just three of the four old essences, or *earths,* as they were now called. They were *vitreous earth,* which gave solidity to matter; *fluid earth,* which gave it liquidity; and *fatty earth* (later called *phlogiston*), which gave it combustibility. These were the old Aristotelian elements of earth, water, and fire. But air was now excluded! By this time, air was thought to be inert and not a part of other materials. Nothing in the old alchemy revealed the role of oxygen in combustion, and not until we understood that role would we be able to replace alchemy with a useful atomic theory of matter.

Of the four essences, fire was the one that would ultimately lead people to understand the role of oxygen in combustion. Scientists made a major breakthrough in the late eighteenth century when they realized that heat was not a part of matter. They dropped the old concept of phlogiston (which was a component of matter) and invented a new alchemical essence, which they called *caloric.* Caloric was imagined to occupy all matter and to flow from hot bodies to cold ones. Caloric was such a compelling concept that it lingered even after the atomic theory of matter replaced the various earths. It was still being used to describe heat when my grandfather was a little boy. Today we know that heat flow is actually a transfer of kinetic energy from molecules in one region to molecules in another. But it's a tribute to the old alchemists that most of us still envision a fluid moving from hot objects to cold ones when we talk about the "flow" of heat. Modern physics aside, our concept of heat still shapes itself around the old and plausible idea of caloric, not jiggling molecules.

The famous French scientist Antoine-Laurent Lavoisier, who was heavily involved in questions of combustion and chemical reaction, gave caloric its name. Today we know that water is formed by the reaction of two chemical elements that were just coming into focus in Lavoisier's time, oxygen and hydrogen. Lavoisier is best known for his work in isolating oxygen, but he also named hydrogen in 1783 when he realized that it makes water when it is burned in oxygen. *Hydrogen* means "maker of water" in Greek. (The element had been recognized long before Lavoisier. The sixteenth-century Swiss alchemist Paracelsus had separated it, although he had confused it with other flammable gases.)

The same year Lavoisier saw the role hydrogen played in combustion (1783) the French began making balloon ascents in Paris. The Montgolfier brothers used hot air in their first manned balloon. But while hot air is easy enough to come by, it is only a litle less dense than the cool air around it. Since hydrogen weighs only a fifteenth as much as air, it offered terrific lifting power in a balloon. The person who championed the use of hydrogen in balloons was Jacques Charles, a French natural philosopher who flew an unmanned hydrogen balloon just before the Montgolfiers' flight and a manned one

Alexandre Charles filling his hydrogen balloon, from the 1897 *Encylcopaedia Britannica*.

only three and a half months after it. He invented a hydrogen generator that worked by mixing huge quantities of sulfuric acid with iron filings.

By this time science in both England and France was heavily dependent upon external observation. Strong lingering aftertastes of the old alchemical essences remained, but alchemical methods were all but forgotten. In France the wedding of the new sciences with technology, however, was on rocky ground. It was acceptable for science to join hands with ballooning, for balloons were the work of sportsmen, but French scientists and French industrialists did not work together.

Things were different in England. The industrialists Boulton, Watt, and Wedgwood, for example, joined with scientists Priestley, Herschel, and Erasmus Darwin in a Birmingham scientific club called the Lunar Society. Science was no sport for these people. It was a means for the pursuit of social reform. They took an interest in balloons not out of any love of flying but for the scientific possibilities balloons offered. For example, in 1784 Watt and Boulton wanted to find out whether the reverberating sound of thunder was the result of repeated claps or of echoes. So they built an unmanned paper balloon, filled it with air and hydrogen, and sent it up with a timed fuse. They hoped to find out whether a single explosion resulted in reverberation. The blast they set off in the sky was indeed a grand and soul-satisfying one, but they did not manage to sort out any pattern of echoes.

The early balloonists thus put the evolving language of chemistry— terms such as *inflammable air* and *phlogiston*—on every tongue. These huge hydrogen-filled spheres dressed the sky in fantastic colors and caught the public's fancy. Balloons ultimately did hurry the process of making chemistry a servant of the Industrial Revolution.[9]

But it was the study of heat and combustion that finally showed us how to break away from the old alchemical essences. Until we under-stood heat we would not be able to sever the last cords that still bound us to earth, fire, and water (and possibly air). And here we face a won-derful contradiction, for we would never find our way to a full-blown atomic theory until we once more learned to articulate it deductively and theoretically. We had to rediscover some of the lost Platonism.

The early nineteenth century finally saw the pendulum swing back toward the viewpoints and methods of the old alchemy. Perhaps the one person who most clearly articulated the need to see with the eye of the mind once again was John Tyndall, an Irish physicist who taught at London's Royal Institution. Tyndall read the new Romantic poets, and he combined their insistence on harnessing the power of the mind with his own remarkable processes of observation. He showed scientists how to live in both worlds.

We begin to see how Tyndall's thinking worked when we read his study of sound. For example, he picks up on a stray observation by an American professor named Le Conte. Le Conte had gone to a lamplit musical party and then written about it in the 1858 *Philosophical Magazine:*

Soon after the music began, I observed that the flame exhibited pulsations that were exactly synchronous with the audible beats. This...phenomenon was very striking, [especially] when the strong notes of the 'cello came in. It was exceedingly interesting...how perfectly even the trills...were reflected on the sheet of flame.

Surrounded as we are by microphones, amplifiers, and oscilloscopes, it is hard to imagine a time when the study of sound wasn't equipped with high-tech electronics. Victorian scientists had nothing of the kind. The observant Le Conte, his mind divided between music and physics, seized on this new means for diagnosing sound. Tyndall had a genius for fusing observation with theory and invention. When he read Le Conte's account he saw vast possibilities within it.

John Tydall, 1820–1893.

Tyndall, born in 1820, had first worked as an engineer designing railway equipment and eventually earned a doctorate in Berlin. The range of subjects he contributed to was vast—heat, light, electricity, glaciers, microbiology, and the Victorian science-religion controversies. His book on sound shimmers with highly honed mechanical means for displaying sound. In it he developed Le Conte's observation into an eighteen-page section titled "Sensitive Naked Flames." Tyndall used flames to measure pitch and tone quality. At one point he read lines that he attributed to Spenser while a flame flickered across a room:

> Her ivory forehead full of bounty brave,
>   Like a broad table did itself dispread;
> For love his lofty triumphs to engrave
>   And write the battles of his great godhead.
>
>   ...and when she spake,
> Sweet words like dropping honey she did shed,
>   And through the pearls and rubies softly brake
> A silver sound, which heavenly music seemed to make.[10]

As he read, he watched how the flame danced to different vowel sounds. The choice of Spenser, by the way, could hardly have been casual, for Spenser was a sixteenth-century poet acknowledged by the Romantics as a kindred soul.

Flames were only one of a score of means that Tyndall used to analyze sound. He reflected light from tuning forks onto moving paper. He gazed at the effects of sound on water jets and shallow sand. There seemed no end to his methods, all so direct and palpable. We are where we are today because people such as Tyndall finally showed us how to be intelligent observers and still bring the deep-seated forces of our subjective intellect to bear upon observation at every step.

Tyndall's contribution thus reached far beyond his vast body of scientific work. He showed how we might return the mind to full participation in the process of observation. And he did so at a time when the scale of scientific measurement was going quite beyond anything previously known. By the end of the nineteenth century, scientific observation had taken on the grand scale of the industrialized world around it. Let us look briefly at an example of just how far experimentation had gone—at the backdrop against which Tyndall was pressing for a return to some of the old Platonist virtues.

Nowhere had scientific measurements been made on as grand a scale as in the field of astronomy. And no astronomer measured on a grander scale than did George Ellery Hale. Hale was born in 1868, just after the American Civil War. As a twenty-four-year-old professor at the University of Chicago, he organized the Yerkes Observatory, and there he built the largest telescope ever to use a conventional refractor lens. It was over three feet in diameter.

The Yerkes telescope was nevertheless something of a dinosaur, since astronomers gave up conventional lenses in favor of focused mirrors after 1900. But Hale was no dinosaur. By 1904 he had convinced Andrew Carnegie to give him $150,000 dollars to set up the Mt. Wilson observatory in California. Hale was downright greedy for high resolution, and straightaway he developed the largest mirror telescope ever built. It was five feet in diameter.

At first he joyfully cried, "With this we'll record...a billion stars!"[11] But by 1918 he was back at Carnegie's door for money. He convinced Carnegie to support a second mirror, this one a hundred inches (more than eight feet) in diameter.

Hale was now only fifty years old, but his health had begun to fail. Though he had to retreat from fieldwork in astronomy, that did not stop him from planning, writing, and organizing. In 1916 he founded

Science Marries into the Family

the National Research Council, and he was now deeply involved in the task of setting America's scientific research agenda.

But one more telescope was on Hale's agenda—a *really* big one. Andrew Carnegie was dead by now, but the Rockefellers gave Hale $6 million for a third mirror—one almost seventeen feet in diameter. That mirror was to become the heart of the Mt. Palomar Observatory, also in California.

Hale's 100-inch (8-foot) telescope at Mount Wilson. The mirror is held in the metal housing at the bottom.

In 1934 the Corning Glass Company tried to make the first rough casting of this seventeen-foot mirror. They cooked a fifty-foot lake of molten glass for six days at twenty-seven hundred degrees Fahrenheit. When they poured it into the mold, with the press watching, the inside of the mold broke up. Nine months later they tried again and succeeded. It took eight more months to cool it down. Grinding it by hand to tolerances of a millionth of an inch took years.

Hale died in 1938, and the Palomar telescope was finally finished ten years later. Until the Russians made a larger one in 1986, it remained the grandest optical telescope. Compare its five-hundred-foot focal length with the eight-inch focal length of a long-distance zoom lens on your camera. Not until we had radio telescopes would we improve on its resolution. Hale's unmatched vision, nerve, and determination have to move us. Just think: Over a fifty-six-year period, from the age of twenty-four until ten years after his death, George Ellery Hale gave us the world's largest telescope not once, but four times.

And so late-nineteenth-century experimental science seemingly paralleled the grand reach of American industrialization. At the same time, the experimentalist John Tyndall very effectively told scientists that they must process fact on a deeply subjective level. The tension between subjective and objective science was itself being expressed on a truly grand scale.

For that reason, we finish with the ultimate display case for the tension between alchemy and Bacon's science of observation—and the ultimate meeting ground of science and technology. It is in the Second Law of Thermodynamics that science reaches its most subjective implications yet remains constrained by rigorous observation. Nowhere is science simultaneously more abstract and more earthbound than in this law.

The Second Law tells us that order naturally moves to disorder—that no spontaneous process can ever be completely undone, that every living thing must sooner or later die, and that even the sun will eventually burn itself out. Mother Goose stated the law very accurately:

Humpty Dumpty sat on a wall.
Humpty Dumpty had a great fall.
All the king's horses and all the king's men,
Couldn't put Humpty together again.

The Second Law is rooted in probability and it says nothing about individual molecules. It comes about because great numbers of molecules have to go from less probable arrangements to more probable ones. Our awareness of things is on a scale much larger than molecules. Our consciousness follows probability to greater disorder. In that sense, the Second Law could not exist without our subjective perceptions of order and disorder.[12]

The Second Law has been called time's arrow because we experience events only in that one direction. In his paraphrase of Psalm 58, the seventeenth-century writer Isaac Watts caught this directionality of time with these haunting words:

Time, what an empty vapor 'tis;
        and days how swift they are.
Swift as an Indian arrow flies;
        or like a shooting star.

Yet a closer look tells us that the true state of affairs is not so dreary. The less well-known principle of LeChatelier and Braun limits the Second Law. It says that when natural processes go to greater disorder, they at least summon up resistance to their own completion. The best-known example is chemical reaction: The hotter any flame becomes, the less complete combustion will be. Instead of racing to completion, burning opposes its own action. Rust likewise tends to cover metals with protective coatings that slow the process of rusting. Spontaneous processes degrade things, but nature inevitably invokes processes that slow degradation. All natural processes resist the Second Law in one way or another. Nature protects us; it slows and inhibits the inevitable, and it grants us time.[13]

In other words, if we remember Shakespeare's wonderful version of the Second Law, "Golden boys and girls all must, / As chimney sweepers come to dust," we must also remember that our coming to dust is muted by the LeChatelier–Braun principle. We *can* grow old gracefully. The Second Law aims time toward disorder and decay, but LeChatelier and Braun tell us that time's arrow is slowed—that it is possible to sustain, and enjoy, some measure of beauty and order along the way.

Bacon's advice that nature "should be discovered, not imagined or invented," so powerful and so far-reaching, changed the course of science and technology utterly. Yet we find that it was dangerous advice, for the human imagination is too necessary a part of all we do. And in the fully articulated Second Law of Thermodynamics, the world that we observe objectively fuses with our subjective substrate.

The Second Law is a place where mathematics, spun in our mind's eye, has been used to make *objective* sense of a set of facts grounded in *subjective* experience. In the end we realize that for us to be good engineers or good scientists, these contradictions must marry and become one within us. We find at last that our mortality is, in the same breath, a cold, objective fact of life and a completely subjective facet of a larger process, which is worthwhile after all.

6

Industrial
Revolution

The Industrial Revolution is an easily misunderstood event. In many people's minds the phrase suggests mass production, assembly lines, and the heavy industry of the late nineteenth century, but these things all came much later. When Arnold Toynbee coined the term *Industrial Revolution,* he applied it to the technology-driven change of British life as it occurred from 1760 to 1840, opening a very large umbrella. Yet even that umbrella still did not cover the first mass production and assembly lines, nor did it encompass our images of modern heavy industry.

Toynbee's dating of the Industrial Revolution starts when its causes were just taking form, and ends when England had become a mature industrial power. He took in the whole saga of the revolution, but within that saga we can identify the Revolution as a much more specific moment in British history. It is the point at which technology suddenly joined hands with radical social and economic changes. In the 1780s Watt's advanced steam engines, Hargreaves' spinning jenny, Cort's improvement of wrought-iron production, and Wilkinson's cylinder-boring mill all came into being. At the same time, economic theoreticians David Hume and Adam Smith were setting forth a new economic and social system.

This convergence of inventions was part and parcel of the other great revolutions of the late eighteenth century—the American Revolution, the French Revolution, and a spate of lesser European revolutions. We have to understand it in the context of those political and social upheavals. In England, social revolution grew out of eighteenth-century

Protestant reform. The Wesleyan movement and the various dissident Protestant groups counted the makers of the Industrial Revolution among their members.

The mid-eighteenth century was marked by worldwide discontent with authoritarianism and with the tyranny of the mercantile economic system. The French kings loved elaborate clocks and clockwork toys—devices that were completely preprogrammed. By the late seventeenth century, they had joined with the other western European nations in a clockwork economic system as well. The mercantile economic equation specified trade balances, such that raw material flowed in, manufactured goods flowed out, and gold flowed in. To make it work, a nation had to increase its labor force while it minimized the wants and needs of that labor force. Nations needed colonies to provide the raw materials and to provide markets for the manufactured goods.

At the heart of eighteenth-century revolution was the realization on the part of workers that they had to lay claim to the fruits of production. On an almost visceral level, the citizens of Europe, England, and the American colonies saw that, to lay that claim, they would have to gain control of their technologies.

In England, however, the citizenry circumvented political revolution and seized technological power directly. Protestant reformers, far from the center of political power in London, established a new power-generation industry. As early as 1712 an Anabaptist blacksmith from Devonshire in southwest England developed and patented the first practical steam engine. He was Thomas Newcomen, and throughout the eighteenth century English commoners like Newcomen seized power (in the most literal sense of the word) and harnessed technology as the means to better their lot.

Earlier technologies had not been driven by that kind of revolutionary purpose. Now, suddenly, technologists were consciously working to divert manufactured goods back to the people who made them. One does not always think of the Industrial Revolution as social reform. Clearly, by the mid-nineteenth century industrialists had become a new ruling class—the new oppressors—but that was later, and part of another story altogether.

As noted in chapter 1, technology is the authentic means by which we change the character of our lives. It was rightly the heart and soul of the social revolution that swept the entire Western world in the late eighteenth century. So it should be no surprise that an early manifestation of this form of revolution was the appearance of a new class of technical encyclopedias. Encyclopedias date back to second-century Rome, but Ephraim Chambers' English encyclopedia broke new ground in 1728. Its title, typical of the time, sounds more like a table of contents: *Cyclopaedia; or a Universal Dictionary of Arts and Sciences, Containing an Explication of the Terms...in the Several Arts, both Liberal and Mechanical,...etc.* The new idea here is *mechanical arts.* Earlier encyclopedias had never been that down-to-earth.

The French arranged to publish Chambers' encyclopedia in 1745. But after a fight with the English translator, they decided to develop a greatly expanded French version instead. By 1747 Denis Diderot had assumed leadership of the whole project except for the mathematical parts, which were handled by the mathematician D'Alembert.

Diderot added fire to the project. He was briefly jailed in 1749 for his liberal views, and when the first two volumes were published in 1751, he was attacked by Jesuit authorities. The work was now titled *Encyclopedia, or a Systematic Dictionary of Science, Arts, and the Trades,* and it was on its way to becoming a twenty-eight-volume treatise on human affairs.[1]

The problem was that Diderot and the other writers were *rationalists.* The *Dictionnaire,* as it was called, laid bare the workings of the known world as no one had ever tried to do. It boldly told commoners that they could know what only kings, emperors, and their lieutenants were supposed to know. It suggested that anyone should have access to rational truth. In that sense, it was a profoundly revolutionary document, and the French government did not miss the point. They twice tried to suppress the *Dictionnaire,* yet the work was completed in 1771, a little more than a decade before the French Revolution, which it was subtly fueling.

The *Dictionnaire* nurtured revolution both by including the trades along with the arts and sciences and by recognizing the intimate link between technology and culture. And oh, how gracefully it did that job! Its beautiful plates show, in wonderful detail, how tanning, printing, metal founding, and all the other production of the period was done.

But the *Dictionnaire* was published on the eve of the English Industrial Revolution, as well as the French Revolution. The trades Diderot described were just about to undergo a complete transformation. The technical descriptions ultimately served the history of technology better than they served technology itself. The irony is that this radical set of books left us with a fine record of techniques that had been perfected, with little real alteration, since the Middle Ages. The very revolution the *Dictionnaire* was fomenting would soon replace most of the techniques that it detailed with such loving care.

Wagon wheel making, as depicted in Diderot's *Dictionnaire*.

The way most French technology actually worked was quite different from the down-to-earth impulses of the *Dictionnaire*. Let us take a moment to savor an example. Around 1750 Louis XV's mistress, Madame de Pompadour, wanted a water supply for her château at Crécy, and the job of providing one fell to the noted French mathematician Antoine de Parcieux.

Why a mathematician? Because eighteenth-century rationalism was just catching up with the medieval waterwheel in 1750. Power-producing waterwheels took many forms, but the field had narrowed to two types by 1750—overshot and undershot wheels. A high-speed stream directed beneath an undershot wheel drove its rotation. A slow-moving stream entered above an overshot wheel and the weight of water caught in its blades forced the wheel to turn. The eighteenth century was marked by a tremendous drive to harness more power for industrial use. It was important to know which kind of waterwheel would give more power.

Antione de Parcieux correctly saw that the overshot wheel would produce more power for the pumps at Crécy. But his choice turned out to be fortuitous, because his calculations were in error and his experiments were crude.[2] Isaac Newton had provided the intellectual apparatus for analyzing waterwheels in 1687. The names of the people who studied waterwheels form a roll call of the great mathematicians and scientists in the mid-eighteenth century. They included even Leonhard Euler and the Bernoullis. Finally, the English engineer John Smeaton put the question to rest. Smeaton was the prototype of eighteenth-century engineering. He designed the first successful Eddystone lighthouse, greatly improved Newcomen's steam engine, and designed windmills. In 1754 his systematic scientific experiments made it clear that overshot wheels were better than undershot ones. About the same time, Euler's son Johann came to the same conclusion using a correct analysis. This was just fifteen years before Watt patented a superior steam engine in 1769. It has been argued that Smeaton slowed the spread of steam power with his fine work on the waterwheel. But it would have taken more than an improved waterwheel to slow the inexorable force of steam power.

Nevertheless, the body of theory these people left reappeared in the nineteenth century in modern power-producing water turbines. Even the new water turbines could not slow the spread of steam power. Water turbines provide a good fraction of the energy we use today by taking it cleanly and cheaply from rivers and waterways. However, since rivers can supply only so much power, the fraction they provide shrinks as total energy consumption (from all sources) keeps increasing.

So the next time you visit Grand Coulee or Hoover Dam, try (if you can) not to think about King Louis XV indulging Madame de Pompadour with an improved water supply for their love nest. Think instead about the entirely different modality of English technology. Nowhere do we more accurately catch that mood than in a small group called the Lunar Society of Birmingham. The Lunar Society met monthly throughout the late years of the eighteenth century and into the nineteenth century.

Revolutionaries have always gathered in small groups, and the revolutions of the late eighteenth century were no exception. They took many forms, but they were all fomented in study groups. These groups invariably got around to a common question: How could science and technology serve society? Before the French Revolution, intellectuals—both men and women—met in salons to talk about scientific and social issues. Ben Franklin set the pattern in the colonies with his American Philosophical Society, Franklin's life was centered both on revolution and on tying scientific knowledge to practical social change.

The Lunar Society took its name from the fact that it met during the full moon. That way, roads were better lit for members who had to travel at night. It numbered only about a dozen people at any time, but what a dozen they were! The heart of the society was Matthew Boulton (the industrialist who built Watt's engines). Other members included Erasmus Darwin (a well-known physician and writer, and Charles Darwin's grandfather), Joseph Priestley (a rebellious cleric and scientist, famous for isolating oxygen), and Josiah Wedgwood (who did not just make fine tableware but was also dedicated to improving the quality of everyday life; and he was Charles Darwin's other grandfather). The membership list goes on: astronomer William Herschel (who discovered the planet Uranus but was also a noted organist) and John Smeaton (whom we have just met).

Can you imagine being in a room with these makers of the Industrial Revolution who were genuinely seeking means for improving their world? Historian Jacob Bronowski looks at the Lunar Society and says, "What ran through it was a simple faith: the good life is more than material decency, but the good life must be based on material decency." It comes as a jolt to see dedicated capitalists forming a revolutionary cabal. But in 1785 capitalism *was* revolution. When these eighteenth-century intellectuals and industrialists consciously joined forces, it was because they wanted to shape a decent life for everyone.[3]

Forty years later, in 1825, John Nicholson published a work called *The Operative Mechanic and British Machinist*. There we find the impetus of the Lunar Society still alive and well, even though its members had by now all passed on. Today, machinists' handbooks are collections of hard information about thread sizes, standard metal thicknesses, and working tolerances. They deal in the specifics of technology and say little about its broad sweep. But Nicholson's work emerged from the smoke of the Industrial Revolution. A dazzling profusion of new machines had come into being, and this eight-hundred-page, two-volume compendium sets out to explain them all. This is no shop guide to nut and bolt selection. The mechanics and machinists he addresses are the engineering designers of 1825.

Thumb through its plates, and look at the power transmission devices, systematic studies of animal and human power output, hardware for harnessing and controlling waterwheels, complex windmill systems, flour mills, steam engines, papermaking, printing, weaving, pumping. The book says almost nothing about how to make these devices. Rather, it speaks to people who already knew the machine tools and processes that built this glorious inventory of machines. But the most telling part of this book is the inscription at the front. Nicholson says to his readers,

> In an age like the present, when the rich and the powerful identify their interests with the welfare of the poor and uninformed, when the wise and good combine in furthering the diffusion of sound principles and useful knowledge among those who constitute the

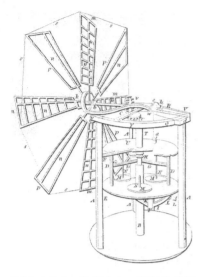

Nicholson's *The Operative Mechanic and British Machinist* shows us how to build a windmill.

most important, though hitherto the most neglected, portion of the community, there is not one who can view the future in the past but must anticipate with such data before him, a change as brilliant in its effects, as it is honorable to those who are engaged in promoting it.[4]

Nicholson expresses two essential sentiments of the Industrial Revolution here. One is that technologists are responsible for improving the lot of the poor. The other is that good work rewards the technologist who does it. People who have read Charles Dickens have trouble understanding that this sort of high principle drove those who created the Industrial Revolution. Dickens was still in grade school in 1825, but the engines of greed were already tearing the fabric of this idealism. People sitting in offices, away from the noise and smoke, were creating the workers' hell that Dickens would begin describing fifteen years later. But what we see here is the last of a breed who really did, in Nicholson's words, "identify their interests with the welfare of the poor."

The Industrial Revolution was driven first by idealism. We see grace and form in Nicholson's drawings of machines. It is beauty that reflects a high purpose we have almost forgotten today. We think of the Industrial Revolution and envision not social reform but steam engines and railways.

It is important to see these machines in the context of the times and to take stock of the limited extent to which the steam engine was at the heart of it, even though it certainly was England's great gift to the eighteenth century. When Boulton and Watt began selling Watt's engines in 1776, steam engines had been around for seventy years. Almost six hundred of them had been built. Watt made improvements that quadrupled their efficiency. His first engines put out only about six horsepower—not much more than the first Newcomen engines—but they were smaller, and they used far less coal. In less than twenty years he had increased the output to as much as 190 horsepower.

In the eighteenth century 190 horsepower would by no means fit under the hood of a car. Those early engines were enormous. An old Newcomen engine had a cylinder anywhere from two to ten feet in diameter, and the engine itself was a two-story structure. Watt's engines were more compact, but their cylinders were still between one and a half and five feet in diameter. And, good as they were, Watt's engines did not dominate power production. By the end of the century, just over two thousand steam engines had been built in England, and fewer than five hundred of them were Watt engines.

Actually, steam engines *never* became a major power source in the eighteenth century. Most power still came from waterwheels and wind-mills. Steam-engine factories never produced more than a few hundred horsepower total per year. But steam power picked up those specialized tasks that were absolutely essential for the rest of the Industrial Revolu-tion to take place—such as pumping water out of mines so the needed coal and metals could be reached.

By 1800 the total power capacity of all the steam engines ever built was about the same as one larger diesel engine today. While they did not change the English countryside overnight, they were the stalking horse of the greatest revolution the world had ever seen—agents of changes that far outstripped anything their makers had ever imagined.

Another agent of change was the sudden explosion of a transporta-tion system for hauling raw and manufactured goods. The English never had been serious road builders. They had done little to surpass their old Roman road system in fifteen hundred years. But now outly-ing English traders saw the potential in moving their own goods about, and during the late 1700s they began creating a very effective canal system.

They also developed a railway system, but not in the sense we think of it. The steam locomotive would not be invented until much later. Instead those merchants built horse-drawn trains. As canals became the major means for hauling goods cross-country, portaging inevitably had to be done between canals. Since roads could not stand up to heavy wheeled vehicles, the English developed horse-drawn railways for portage. That idea had, in turn, come out of the mines, where tramways were used to move coal and ore. When the steam locomotive was finally invented, the technology of building railways had been well developed in the mines and around canals.

Engineers also knew a lot about the loads horses could pull. At a slow walk, a horse could pull almost thirty tons through a canal, but only seven tons on a railway. As the horse sped to a trot, water resistance became so great that it could pull almost nothing. But on a railway, it could pull just as much at a trot as at a walk. A horse could move more goods on a canal, but when speed was needed, it did far better on rails.

The Cornish builder Richard Trevethick built the first steam locomotive in 1804, and railroad speeds increased rapidly from then on. Water resistance made canals quite useless at the speed of a train. So from the early nineteenth century until the modern automobile, railways dominated English transportation.

The land locomotive—the early steam car—made a valiant try during those years. But it was easier to develop a rail system than a road system to support such heavy vehicles. So many factors were at play in that brief eighty-year period! And out of a gaggle of opposing means, the railway emerged to dominate transportation for about a century. But even here, the forces of revolution remained in play.

Consider for a moment the word you might well have used for a train when you were a child—the word *choo-choo*. *Choo-choo* was the noise made by steam leaving the cylinders down by the wheels. If you have never seen it in real life, you have seen it in movies. A conductor shouts, "All aboard," steam gushes about the wheels, and the train starts to move. That *choo-choo* sound reflects two ideas that converged around 1800, after steam engines had been in use for a hundred years. One was the idea of running steam engines at high pressure; the other was using them for transportation.

The first steam car was made by the French military engineer Nicolas-Joseph Cugnot in 1769. Since steam engines were huge two-story structures, it's not surprising that Cugnot's car was a big brute. It carried four people at about two miles per hour. It was meant to pull field artillery, but it was not really practical. In 1784 William Murdoch, who worked for Watt, used a Watt engine to produce a better car, one that was lighter and faster. The problem was, Watt disliked the idea of using steam for transportation. He patented the idea so he could put it on ice.

Watt disliked high-pressure steam as well, and that's the other thing needed to make a vehicle. Early steam power plants all depended on condensing steam to create a vacuum. They worked by sucking the piston into the cylinder instead of by pushing it out. Low-pressure steam takes up space, and that made engines large. When the pressure reached fifty or a hundred pounds per square inch, the engines could be made a lot smaller. Watt wanted no part of that game, because high-pressure steam could be dangerous.

The improved boring and machining equipment of the late eighteenth century finally made high-pressure engines realistic. Trevithick and the American Oliver

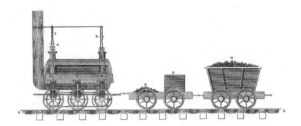

An unspeicifed early railway locomotive, shown in the 1832 *Edinburgh Encyclopaedia*.

Evans both made good high-pressure, noncondensing engines just after 1800. Their engines had small, well-machined cylinders that would fit in a vehicle. And instead of condensing steam to create a vacuum, they discharged spent steam into the atmosphere, making that *choo-choo* sound.

Trevithick and Evans both used their steam engines to drive cars— without much success. But then Trevithick saw that steam could replace the horses that drew carts on England's rail system. He was clever in selling the idea. After he made the first successful locomotive in 1804, he built a little demonstration railroad in London in 1808. It was a sort of carnival ride, with the train (called the "Catch-me-who-can") running on a loop of track at a swift twelve miles per hour. From then on, steam trains really caught on. The high-pressure steam engine opened up western America. That familiar *choo-choo* sound of spent high-pressure steam is a sound that tells how two good ideas finally came together.[6]

But we are now getting ahead of our story, for transportation was a significant part of the Industrial Revolution as it was about to reveal itself here in America. Part armed rebellion, part technological revolution, and part spiritual regeneration, the creation of the United States was unlike anything the world had ever seen.

7

Inventing America

merica was not discovered, it was invented. Its name was invented; its machines were invented; its way of life was invented. America sprang from the minds of that unlikely breed of people who were able to pack up a few belongings and step into a great unknown. That step into the expanse of a new continent unleashed astonishing creative energy.

America was an adventure of the mind. The land seemed to reach into infinity, and minds opened to fill it. The colonists had limited recourse to the European intellectual mainstream. They were poorly equipped, but they were freedom-driven and freedom-shaped. They were free of method and free of tradition. They were free to create a new life.

Colonial technology was so molded by the imperative to be free that it is hard to talk about it without being drawn into that infectious drive. You cannot just report it; you have to celebrate it. As I look back at the early episodes of *The Engines of Our Ingenuity* upon which this chapter is based, it is clear that I too was drawn in. My first impulse in reworking this material for print was to tone it down and mute my enthusiasm. In the end I did not do so. History gives us too few moments with such verve. Why not go back and be the irrepressible child that America itself once was?

The need to rediscover the childhood of our nation is great. We are drifting into a new sobriety. It was in my generation that we first lost a war. We no longer take our leadership in productivity for granted. We have found that we have a capacity for failure, and that we do not always emerge as the good guys. We have deconstructed our heroes until they seem to be heroes no more.

But they *were* heroes. Any chapter on colonial technology inevitably yields up the names of Jefferson (no mean inventor himself), Fulton (with his thumb in so many pies), and the towering figure of Benjamin Franklin. These people appear here not because they were the only heroes we had, but because they were true paragons of colonial creativity. They represent its full breadth and intensity, but their talents were mirrored in thousands of other creative early Americans—many of whom we shall meet as well.

If this chapter is filled with hyperbole—not just my own, but that of all the characters who speak to us from those early days—I am hardly inclined to apologize for it. The language of our colonial forebears outraced even their dreams, and their dreams were already much larger than harsh colonial life gave them reason to be. Nowhere was the dream so large and the language so grandiose as it was in the Declaration of Independence. When the fifty-six signers asserted that this outback would "assume among the powers of the earth, the separate and equal station to which the Laws of Nature and of Nature's God entitle them," they claimed equality with countries that hardly knew they existed. When they called equality self-evident and liberty an unalienable right, their language outstripped the realities not just of colonial America but of the human condition. These people showed us the way any creative person must lose touch with reality when realism can mandate only what is obvious. It was precisely their ability to see beyond all the obvious limitations that allowed them to invent America.

The first invention was a name for the new land. We named the American continents after the Italian navigator Amerigo Vespucci instead of his countryman Christopher Columbus. But why? Who was Amerigo Vespucci, and what did he do?

He was an Italian merchant, born in 1454 in Florence. He worked for the Medicis, and in the year Columbus made his first voyage, they sent Amerigo to handle the Medici ship-outfitting business in Spain. He helped to outfit Columbus' third voyage. Finally he outfitted his *own* voyage to look for the Indian subcontinent (which had eluded Columbus). He sailed in 1499, seven years after Columbus first landed in the West Indies. Vespucci made trips in 1499 and 1502, and possibly a third one in 1503.

During his first voyage, he explored the northern coast of South

America to well beyond the mouth of the Amazon. He gave Asian place names such as "Gulf of the Ganges," to the things he saw. He also made significant improvements in navigational techniques. On that trip he predicted Earth's circumference to within fifty miles.

But a breakthrough came on Vespucci's second voyage, when he realized that he was looking not at India at all but at an entirely new continent. He verified the fact by following the coast of South America to within four hundred miles of Tierra del Fuego. Columbus may have found the New World, but Vespucci recognized that it *was* a new world.

So who wrote Vespucci's Christian name, Amerigo, on the maps? Was it the king of Spain, our founding fathers, Vespucci himself? It was none of them. An obscure German clergyman and amateur geographer named Waldseemüller belonged to a literary club that published an introduction to cosmology in 1507. In it he wrote about the new land mass Vespucci had explored: "I see no reason why anyone should justly object to calling this part...America, after Amerigo its discoverer, a man of great ability."[1]

The name stuck. Later Waldseemüller thought better of the name, but he had not counted on the power of the new medium of print. There was no undoing what had been set in type. When a second great land mass was found to the north, the names North and South America were applied to both continents. And we're left with an old question: Who really discovered America on behalf of Renaissance Europe? Was it the person who found it or the person who recognized it for what it was?

That riddle dogs all of science. Equally futile arguments rage over the question of who discovered oxygen. Was it Priestley, who first isolated it; Lavoisier, who recognized it as a new substance but failed to identify what the substance was; or Scheele, who got it right before either Priestley or Lavoisier but didn't publish until *after* they had?[2]

It is (as we shall see in chapter 14) a hopeless question. Columbus, Vespucci, Ericson, and that brave Asian who first walked off across the tendril of land that is now the Bering Strait—they all discovered this place. We might as well say with Waldseemüller, "Why *not* call it America?" for that makes as much sense as *Columbia or Ericsona*. (Actually a Chinese explorer named Hoei-Shin almost certainly made it to western Mexico in A.D. 499. Perhaps we should call our land *Hoei-Shinia*.)

Inventing America

Once in the New World, we had only to look around to see the potential in new ways of doing things. For example, the United States and Canada enjoy far more extensive waterways—lakes and rivers—than does western Europe. Native Americans had invented remarkable means for using that resource. They created the *Indian canoe.*

As we look at inventions, every now and then we find an idea that does much more than just catch on. Some ideas are so *right* that people improve them until they fit their purpose perfectly. They reach a state that needs no further improvement. Indian canoes are a fine case in point. When I was a child in Minnesota, we took seriously our state's nickname the "Land of Ten Thousand Lakes." That was not just for the tourists. Minnesota is riddled with lakes, and they all have canoes on them. Nowadays, most of those canoes are made of aluminum or fiberglass with polystyrene floats in the bow and stern. But their form is still very nearly a carbon copy of the Indian birch-bark canoe.

Historians have traced the Indian canoe back as far as they can, but Native Americans kept little of what could be called a written record. Since the canoes were completely biodegradable, we have no archaeological remains either. The oldest information comes from early sketches left by the first European explorers. All we know is that canoe making was perfected a long time ago, and for centuries designs have merely fluctuated about that one near-perfect design.

Canoes are shallow-draft boats with a fine, delicate shape. Their hydrodynamic form has much in common with the Viking ship. One advantage it has over a rowboat is that the paddler faces the direction he travels. Most Indian canoes were small, light, and fast. They carried a few people rapidly up and down rivers and lakes. The Iroquois built thirty-foot-long freight-carrying canoes, capable of carrying eighteen passengers or a ton of merchandise. But even the large boats could be portaged by just three people.

The Indian canoe consisted of a tough, light wooden frame with a skin of bark—usually birch. Sometimes the bark was put on in one piece and pleated to take up slack as it was contoured. Sometimes it was sewn in sections and caulked with spruce gum. The techniques of selecting and preparing materials, sewing, binding, and carving were very sophisticated. Designs varied from tribe to tribe, according to local conditions. But even kayaks in the far north, covered with animal skin instead of bark, reflected the same essential concepts of shape and propulsion.[3]

Every once in a while human ingenuity finds a dead end in functional perfection. That is far from true of today's computers or telephones. But it *is* true of stringed instruments, silverware, and lead pencils. And it seems to have been true of that distinctly American icon, the canoe.

The earliest settlers readily adopted the Indian canoe; it was too good not to adopt. However, the earliest settlers did not generally take up native technologies, but instead imposed the technologies of the Old World on this new land. We saw in chapter 1 how this tendency would change as we moved west during the nineteenth century. But in the beginning the game was to re-create Europe as fast as humanly possible. And that meant colonists would need iron, a material the Indians had hardly touched.

Settlers had found iron ore on an island off the North Carolina coast as early as 1585, twenty-two years before the founding of Jamestown and thirty-five years before the Pilgrims landed. The iron was too inaccessible to mine, but Jamestown settlers found ore on the mainland in 1608, the year after they arrived. John Smith shipped several barrels of ore back to England. The East India Company found that the ore yielded top-quality iron, so in 1619 they sent ironworkers to set up smelting operations. Three years later, just as a settlement of twenty-five people was starting to smelt iron near present-day Richmond, Indians massacred them and destroyed their furnaces. Virginia didn't even try to get back into the iron-smelting business, and colonists turned their attention to the Plymouth Colony in Massachusetts.

In 1644, only twenty-four years after the Pilgrims landed, John Winthrop set out to build bog-iron smelters on two sites—one in Braintree, south of Boston, the other just north of Boston on the Saugus River. The Saugus Iron Works operated until 1668, when a labor shortage put it out of business.

The Saugus works was an integrated facility. It had a dam to provide water power for forging, a smelting furnace, a trip-hammer forge, and a rolling/slitting mill. The name given to the factory was Hammersmith, and it produced two kinds of iron. *Cast iron* was poured directly in molds to produce an end product. The other form was cast into "pigs"—large lumps that could either be remelted and cast later or be made into *wrought iron*. To make wrought iron the pig is melted at a high temperature to reduce its carbon content. Then it is forged to refine its grain structure. The result is both tough and strong.

Now, what do you suppose the primary product of the Saugus works was? What do people need when they're trying to build cities from scratch? They need lots of *nails*. The Saugus smiths milled most of their wrought iron into thin strips, then slit those strips into small square rods and sold them to householders. It was up to the user to cut the square rods into short lengths and use small dies to shape points and heads. That kind of nail production was rare in Europe, but our needs were not European needs. New wooden buildings were our first order of business in the seventeenth century. So the Saugus Iron Works was not simply an efficient factory but a visionary response to a basic need.

Today you can visit the rebuilt Saugus works just off Highway 1. The blast furnace is working again. The waterwheel drives the hammer forge and rolling mill. And while you are wondering how they set up such complex equipment so quickly, a tour guide will hammer out a square nail for you right there, where it all began.

Getting new European technologies was a problem, however. In Chapter 6 we saw that Europe's mercantile economic system opposed allowing high technology into the colonies. In the 1670s William Berkeley, governor of Virginia, wrote something that to us sounds astonishing:

> I thank God, there are no free schools nor printing [in Virginia]; for learning has brought disobedience, and heresy…and printing has divulged them, and libels against the best government. God keep us from both.[4]

Not until 1730 did a Virginia governor let a printer set up shop in Williamsburg. Almost immediately a local poet published an ode in praise of the governor, whose name was Gooch. The poetry was wonderful eighteenth-century schlock:

> Truth, Justice, Vertue, be persu'd
> Arts flourish, Peace shall crown the Plains,
> Where GOOCH administers, AUGUSTUS reigns.[5]

It was a battle, but we were determined to establish technological independence. Early American universities studied the new English "fire engines" (as the old Newcomen steam engines were called) in their natural-philosophy courses. American intellectuals kept a close eye on

the technological revolution that was sweeping England. In 1760 young John Adams wrote in his diary that he was struggling to understand the English fire engines. Jefferson studied them at William and Mary College.

We had no need for pumping engines at first. But surface deposits of iron ran out, and scarcer metals, such as copper, could seldom be found on the surface at all. We had to start digging down into the water table. John Schuyler's copper mine near Passaic, New Jersey, was shut down by flooding in 1748, so Schuyler wrote to the English engine maker Jonathan Hornblower. For a thousand pounds Hornblower agreed to ship him a fire engine and a crew of mechanics to set it up. The engine arrived five years later, in 1753, along with Hornblower's son Josiah and his crew.

Josiah got the machine up and running two years after that, so Schuyler hired him to stay on and run the engine and the mine. The engine did well enough for five years. Then it was badly damaged in a fire. Josiah got it running again, but only until another fire ruined it in 1768. This time it stayed ruined through the American Revolution. An aging Josiah Hornblower made another repair in 1793, and this time the old engine kept pumping well into the nineteenth century.

In the long run America could not and would not be built with off-the-shelf English technology. We were starting to build our own engines even before the Revolutionary War. Before Hornblower repaired Schuyler's engine the second time, it had been surpassed not only by better English engines, but by early American designs as well. By 1793 it was already an antiquated tourist attraction.[6]

The real value of Schuyler's tenacity was that it pointed the way to other developers. Colonial intellectuals went to see the engine. Steam power had been a school exercise for Jefferson and Adams. Schuyler's checkered business turned that dinosaur of an engine into a glimpse of America's future.

It should be no surprise that one of the early visitors to Schuyler's engine was Ben Franklin. The French economist Turgot had said of Franklin: "He snatched lightning from the skies and the sceptre from tyrants." That is about as close as one can get to putting Franklin in a nutshell. Franklin's breadth and intelligence were dazzling by any measure. His life exemplified the progress of the American Revolution, from intellectual adventurism in the early eighteenth century all the way to an established new nation.

Franklin devoted his life to understanding and harnessing the forces around him, whether those forces were steam engines, lightning, or political economics. He was only twenty-three when he focused his omnivorous curiosity upon heat radiation. At the time he belonged to a scientific club that would eventually become the American Philosophical Society. Franklin contrived an experiment, and another member helped him. One sunny winter day they laid several colored cloth patches and one pane of glass out on the snow. Then they noted how deeply each melted into the snow. The white cloth hardly sank at all. The darker each patch was, the deeper it sank. The black cloth and the glass pane sank deepest.

Franklin published the results long after he did the experiment. When he did publish, he added a twist: He had also focused a burning glass on both white and black paper. The white paper absorbed less heat, so it took longer to catch fire than the dark paper. Those were results that would not be explained until the twentieth century.[7]

The explanation goes like this: Black cloth is black because it absorbs light. White cloth reflects light. Heat radiation is a lot like light, but not entirely so. Some surfaces respond differently to light than they do to heat. Take skin, for example. Equatorial peoples have dark skins because the pigment protects them from sunburn. You might think that dark skin would absorb heat and make life very unpleasant in hot countries. However, although dark skin absorbs light, less than half the energy carried in sunlight is visible. Most of it is infrared heat, and dark skin *reflects heat* just the way light skin does. If it didn't, Nigerians would all have to live in Stockholm.

Ben Franklin's pane of glass made a similar point. Light goes right through glass. But glass absorbs the heat radiation that is carried in sunlight—it neither transmits nor reflects it. And in that sense, glass is opaque and black under infrared (or heat) radiation.

The young Franklin concluded his study by saying: "What signifies Philosophy that it does not apply to some Use?" He went on to say what his tests suggested about dressing for cold and warm climates. Years later, an aging Franklin watched an early balloon ascent in Paris. The man next to him said, "What good is it?" and Franklin replied with his now well-worn remark, "What good is a newborn baby?"—

which was also a subtle answer to his own earlier question about the value of philosophy.

Early-twentieth-century physicists would show us the newborn baby in the young Franklin's experiments. His simple tests were some of the earliest to begin exposing the way energy is distributed in wavelength. A century and a half later the problem of understanding that behavior would be the stumbling block that eventually drove physicists to create the new quantum theory. Their value lay far beyond the obvious utility of choosing the right clothes for a sunny day.

It would, however, be a mistake to portray colonial America as a new scientific and technological force. The genius and insight of Franklin's work in radiant heat transfer is clear in hindsight, but at the time it was separate from any body of surrounding scholars. Only Franklin's remarkable work on electricity, first published in 1751, found its way into the European academies.

So we remained in the shadow of Europe with far to go. Look, for example, at one grand moment in 1776—the moment when David Bushnell is credited with inventing the submarine. Bushnell built a one-man, hand-crank-powered submarine called the *Turtle*. He enlisted a soldier to take the *Turtle* under the British ship *Eagle* and fasten a bomb to its copper-clad hull. He couldn't drill through the copper, and he had to let the bomb explode in open water, where it did no damage. But the potential of this new weapon was clear enough.[8]

Bushnell gave up submarines a year later and went after British ships moored at Philadelphia with floating mines. Once again he was more impressive than successful. Colonial composer Francis Hopkinson wrote a song about the panic the exploding mines caused on the British-occupied shore:

> Some fire cried, which some denied,
> > But said the earth had quaked.
> And girls and boys, with hideous noise,
> > Ran through the streets half-naked.

Bushnell has ever since been hailed as the father of the submarine and as a great American technological genius. But historian Alex Rowland points out how well informed about European technologies the colonists were. Bushnell worked on the *Turtle* at Yale University, and the Yale library had the English *Gentleman's Magazine*. And the 1747

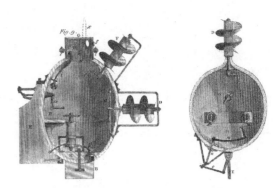

Gentleman's Magazine has a short article with sketches of European ideas as to how submarines might be built. They have the essential features of Bushnell's *Turtle*.

Bushnell's accomplishment was not the invention of the submarine but rather fleshing out a still-skeletal idea. Bushnell was first to put a man under water in combat. The audacity of what he did in a world without any plausible supporting technological infrastructure is stunning.

The mood of colonial America was one of inspired amateurism, and Bushnell was only one of many remarkable exemplars. For me, that mood is gloriously illustrated in an image from the spring of 1786. It is John Fitch's steamboat laboring earnestly up the Delaware River, propelled by an array of Indian-canoe paddles. Twenty-one years before Fulton, those paddles boldly proclaimed Fitch's amateur but functional freedom from any canon of engineering design.

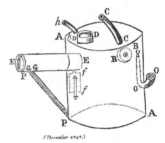

Top: Bushnell's submarine, the *Turtle,* from the 1832 *Edinburgh Encyclopaedia.*
Bottom: Schematic diagram of a submarine, from the 1747 *Gentleman's Magazine,* probably based on a design by the Dutch-English scientist Cornelius Drebbel.

To understand colonial technology and invention we need to understand the intensity of the colonial impulse to be free. The much-used word *freedom* encompassed more than just political independence from England. It included *cultural* freedom from Europe. America's first notable poet, Joel Barlow, repeatedly asserted our cultural independence. He brashly called America a "theatre for the display of merit of every kind."[9] Sometimes this impulse toward freedom was downright arrogant. A typical anonymous Revolutionary War song, set to the skirl of fife and drum, ends with the lines "And we'll march up the Heav'nly streets, / And ground our arms at Jesus' feet." Other times it resulted in gentler expressions of the same sentiment. Francis Hopkinson gave us a widely sung, lilting melody with the title "My Days Have Been So Wondrous Free." But always present was a direct, innocent, homemade, and somehow completely engaging sensibility. It captures our imagination.

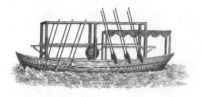

Fitch's first steamboat, 1786, from the 1832 *Edinburgh Encyclopaedia*.

It is strong, affecting, and (most important) completely amateur.

Again and again the mood of Revolutionary America touches us with that direct, simple, but brilliant intensity. Historian Kenneth Clark visited Jefferson's Monticello and said:

> he had to invent a great deal of it himself.... Doors that open as one approaches them, a clock that tells the days of the week, a bed so placed that one gets out of it into either of two rooms, all this suggests the quirky ingenuity of a creative man working alone outside any accepted body of tradition.[10]

We find self-taught Ben Franklin giving us basic insights into the nature of electricity. We find a small band of homegrown intellectuals inventing a new kind of government of, by, and for the people.

The engineering of this new land exhibited the mind-set of people who knew they could do whatever they wanted to do. They knew they could do it better than, and without reference to, what had been done before. Whether it was designing the perfect capital city, building the Erie Canal, or marching their armies "up the Heav'nly street," they knew nothing was beyond them.

The worm of self-doubt afflicts so much we do. We have been to the mountaintop of technological accomplishment. Edison, Ford, and Bell have come and gone, leaving us to feel disconnected from those embryonic years of unreasonable confidence. That worm did not eat into the heart of the people who built this country. With clear, childlike self-assurance, those people quite simply did do the impossible. They could ring the bell that would announce our separation from the most powerful nation on earth.

Now *there* is an evocative technology: bell making. The ghostly Shropshire Lad of Houseman's poem hears the bells that once called his love to church to wed him, then called her to his funeral. He finally screams, "Oh noisy bells, be dumb! I hear you."

Edgar Allan Poe celebrates the "tintinnabulation that so musically wells from the bells bells bells." From sleigh bells and alarm bells to wedding bells, bells signal the turnings and changes of our lives, from the first hour of school to hailing our new nation.

Our Liberty Bell represents a microcosm of political and technological history. Its making was also a kind of laboratory for both politics and technology. The technology of bell making pushes foundrymen to their limits. Bells took their familiar modern form in the thirteenth century—a flattish top, concave sides, and a somewhat thickened lip—but they have always been hard to make. For one thing, the shape is hard to cast. Bronze of 77 percent copper and 23 percent tin seems to be the best alloy. That mix strikes a delicate compromise between tone quality and brittleness.

The bronze Liberty Bell is three feet high, and it weighs one ton. It was made long before the American Revolution. Pennsylvania ordered it from an English bell foundry in 1752 to celebrate that colony's fiftieth anniversary. Even then the inscription from Leviticus inscribed on the bell foreshadowed the role it would later play: "Proclaim liberty throughout all the land unto all the inhabitants thereof."

The English bell cracked even before it was put in its tower, so a Philadelphia foundry undertook to recast it. They tried to make it less brittle by using extra copper, but the bell had a dull sound. So they put the tin back in and cast the third and final bell. It first rang in 1757 at a meeting held by the legislature to create a list of grievances for Ben Franklin to take to England. After that, it heralded the great events of the Revolution. It was muffled to toll a death knell for English taxation; it called rebel meetings; it celebrated victories.

But it did not ring on July 4, 1776. That was merely an intended date. The actual Declaration of Independence was delayed until July 8. Then the bell did indeed ring out, and it went on signaling the great events of our land until 1835. Finally, as it tolled the death of Chief Justice John Marshall, it cracked. It was then seventy-eight years old.

The Liberty Bell had been a true creature of trial and error. It started out English and saw us through to an established America. It followed an odyssey of political change. In the end, it became the perfect national symbol for America.

The quality of invention now began rapidly to shift toward more viable machines. The first celebrated inventor in the new nation was Oliver Evans, a millwright born in 1750 who had one eye on the new medium of steam power. Evans drew on an idea that had predated steam power. As seventeenth-century inventors searched for a new power source, they had experimented not only with steam and air, but with gunpowder as well. Explosives would eventually serve rocketry, but they proved futile for stationary power production. Still, the explosive force that drove a musket ball faster than the eye could follow it kept tempting imaginations even after we had steam engines.

Early steam engines had very little in common with muskets. As noted earlier, they produced power when steam condensed in their huge cylinders and sucked the piston in. They all operated at very low pressures. As a young man, Evans had amused himself with a neat little experiment. He put some water in a gun barrel, corked it tightly, and then heated it until the cork blew out. "Why not make steam engines like that?" said Evans, and he did. The kinship of his steam engines with muskets was quite clear. Their small high-pressure cylinders suggested gun barrels.

He sacrificed some efficiency by discharging spent steam instead of further reducing its pressure in a condenser, as James Watt had done. But his engines were light and they performed well. They were naturally suited to America's need for transportation. Evans spent years trying to find backing for some sort of steam-powered vehicle. It was 1805 before he finally contracted with the city of Philadelphia to build a dredge for their harbor. He closed himself in his workshop while neighbors wondered aloud if he, like Noah, was arming himself against the flood.

One day later that summer, the doors of Evans' workshop finally swung open, and out rolled the most remarkable transportation machine *since* Noah's ark. It was a gigantic steam-powered behemoth that he called an *oructor amphibolos*—Latin for "amphibious dredge." This strange and awesome machine could have lumbered straight off the set of a Mad Max movie. It rolled down the streets, around Centre Square, and into the Schuylkill River, where it chuffed about, dredging mud.[11]

Evans sensed the need for powered transportation, and in one stroke he had made our first horseless carriage, and he had invented a steamboat as well. During the next decade powered transportation began in earnest. Quite suddenly Robert Fulton's steamboat and railroad trains (driven by high-pressure engines) would start carrying us across this sprawling, inaccessible land.

But it would not be steam power alone that would take us into the vast interior of the American continent. Consider a scene from the 1820s. A storm rises over central New York, and a ferryman trudges beside his mule, hauling a barge through the Erie Canal. He sings:

Oh the Erie is a-rising
And the liquor is getting low
And I scarcely think
We'll get a drink
'Till we get to Buffalo.

The Erie Canal is deeply grooved in our national awareness, and it was a marvel. Four of the Great Lakes—Superior, Michigan, Huron, and Erie—all lie above Niagara Falls, and they form a huge inland waterway with access to thousands of miles of shoreline. The waterway touches Minnesota, Wisconsin, Michigan, Indiana, Illinois, Ohio, and Pennsylvania, as well as New York. For our new country to be joined together, East Coast commerce had to gain access to this waterway.

But the inland port of Buffalo, New York, at the eastern end of Lake Erie, is 363 miles from Albany on the Hudson River. Worse than that, Lake Erie lies 568 feet above the Hudson River. Connecting the two ends with a canal was no routine task.

In 1801 Thomas Jefferson appointed a Swiss emigrant, Albert Gallatin, as secretary of the treasury. In 1808 Gallatin presented a proposal to build a giant network of canals, including one between Lake Erie and the Hudson River. In 1810 the mayor of New York City, De Witt Clinton, picked up the idea. His support for the project got him elected governor of New York by a landslide in 1817. Construction of what was to be, by far, the longest canal ever built was ceremoniously begun on the Fourth of July that year.

The task took eight years and seven million dollars to complete. It required eighty-three locks, an 802-foot aqueduct to carry shipping over the Mohawk River, and countless other innovations. Yet the job was done by four principal engineers who had never even *seen* a canal. Like most early American technology, it was the work of amateurs whose zeal and self-assurance took them where angels would fear to tread.

The effect of the Erie Canal on this country was stunning. Cargo that cost $100 a ton and took two weeks to carry by road could now be moved at $10 a ton in three and a half days. Horses and mules drew barges through the canal in end-to-end fifteen-mile shifts. And the ferryman sang the familiar song:

> I've got an old mule, her name is Sal,
> Fifteen miles on the Erie Canal.
> She's a good old worker and a good
> old pal,
> Fifteen miles on the Erie Canal.

Entering the Erie Canal at Troy, New York, from *America Illustrated*, 1882.

The canal fulfilled one of Thomas Jefferson's dreams. Its creation was a task that should have been beyond the engineers who built it, but they simply did not appreciate that fact.[12]

And yet, for all the Erie Canal achieved, it lay far from the eastern Pennsylvania town of Pittsburgh. An account of Pittsburgh in 1816, just before work on the Erie Canal was started, provides a pretty startling window on early American history. The War of 1812 had ended two years before, and our young nation was in surprisingly good shape. Pittsburgh reflected the energy of our new country. Tucked into the western slopes of the Allegheny Mountains, it was centered in the western Pennsylvania coal fields, far from America's population centers on the Atlantic coast.

It costs less to take iron to coal for smelting than to bring coal to iron. So Pittsburgh became our major iron producer soon after the first western Pennsylvania blast furnace was set up in 1790. It became our major glass producer too, because glassmaking also requires a lot of heat. Between 1810 and 1820 Pittsburgh's population mushroomed from forty-seven hundred to more than seven thousand. By 1816 the town had acquired three newspapers, nine churches, three theaters, a piano maker, five glass factories, three textile mills, a steam-engine factory, four thousand tons of iron processing per year, two rolling mills, most of our nail production, and (no surprise) a notorious air pollution problem.

Despite Pittsburgh's lack of access to the East Coast, it sits where the Allegheny and Monongahela Rivers join to form the Ohio River. And the Ohio connects to the ocean at New Orleans, over a thousand miles away. Robert Fulton's steamboat patent was only seven years old in 1816; yet, in that year, this inland city launched three of these gigantic boats to link itself to the sea. And they were not its first; another boat, made two years earlier in Pittsburgh and bearing the unfortunate name of *Vesuvius*, burned in New Orleans in 1816.

These words from an article in the September 3, 1816, issue of the *Pittsburgh Gazette* say much about the mood of the place:

> Those who first cross the Atlantic in a steam-boat will be entitled to a great portion of applause. In a few years we expect such trips will be common...and bold will they be who first make a passage to Europe in a steam boat.

(In fact, the first transatlantic steamboat crossing was made, with the help of some sail, just three years later, in 1819.) The article ends with a quotation from Homer: "Bold was the man, the first who dared to brave,... in fragile bark, the wild perfidious wave." And we read, yet again, the imprint of a developing civilization—healthy, adventurous technologies driven by awe, excitement, and (maybe most important) that perfectly implausible self-confidence.[13]

The story of Pittsburgh once again calls up the name of steamboat builder Robert Fulton. It was just before 1816 that Fulton built his last and most audacious steamboat. When the War of 1812 began, he pitched in to design a really remarkable steam warship. To understand what he did, we need first to look at the Achilles' heel of a steamboat—the paddle wheel. One well-placed shot and *crunch*, the boat was stopped.

Every warship since the Union *Monitor* has solved that problem with submerged propellers. But in 1812 boat propellers had been around for only two decades, and they were still primitive. Paddle wheels were almost the only practical means for propelling a steamboat. The next year Fulton unveiled a ship that had an absolutely protected paddle wheel. It had two hulls, side by side, with the paddle wheel *in the middle*, out of harm's way. He had created a catamaran, 150 feet long and 60 feet wide with a 14-foot slot down the center.[14]

People had trouble naming this strange boat. Fulton called it *Demologos*, meaning the "word of the people." But the navy called it, variously, the *Fulton Steam Frigate*, the *Steam Battery*, and *Fulton the First*. In any event, its double keel was laid in June 1814, and it was launched that October. Four months later the war ended and Robert Fulton died. He was only fifty years old.

The Navy went on to finish the ship, shake it down, and correct a few deficiencies. They were clearly pleased with Fulton's design. The ship saw peacetime service in the New York Harbor area until one summer day in 1829. That afternoon, a gunner went below to the powder magazine to get gunpowder for the evening salute. He carried a candle with him and managed to set off two and a half kegs of gunpowder. Twenty-four men, one woman, and the ship itself perished in the resulting blast.

Fulton might have rewritten naval history but for two things: one was the end of the War of 1812; the other was the development of screw propellers. The next radical shift in naval warship design would not occur until we built the propeller-driven *Monitor* almost fifty years later. Still, Fulton's ingenious double-hulled steam-powered warship was copied many times over during the nineteenth century, for it really was a deliciously good idea.

The next order of business after the War of 1812 was creating the economic means for living the good life. That went on not only in Pittsburgh but all over the country. We were now mounting our own industrial revolution, and Steven Lubar's catalog of a Smithsonian exhibit, called *Engines of Change,* tells about that revolution.

If we began by copying English machines, we did not do so for long. Too much was different in our vast continent. We had resources England did not have. We had far greater potential for water power. We still had abundant wood, while England had long since eaten up her forests. Water and wood rapidly got us started in a power technology based on wooden waterwheels instead of English steam engines.

Our use of wood, coupled with our sense of freedom, had another side effect. We embraced a kind of technology that was ephemeral, subject to change and adaptation, and smaller in scale than England's iron-built, steam-powered juggernaut. Thomas Jefferson opposed large-scale industry, saying, "Manufacturing breeds lords and Aristocrats, poor men and slaves."

Jefferson fostered transportation systems, and he envisioned a widespread, diversified technology. Oliver Evans had characterized the millwright (the typical American engineer of the time) as a true generalist. He wrote:

> [He] could handle the ax, hammer, and plane with equal skill and precision; he could turn, bore, or forge.... He could calculate the velocities, strength, and power of machines; he could...construct buildings, conduits and water courses.[15]

Our circumstances took our technology in directions England's could not go. We were less specialized. We had better natural resources, freedom of ideas, and freedom of movement. We also had to make better use of a smaller labor pool. In 1851 we displayed our wares at the Great Exhibition in London's Crystal Palace. Suddenly a self-satisfied Europe was jolted to see how far we had come. *Punch* magazine gave backhanded praise in a parody of "Yankee Doodle":

Yankee Doodle sent to town,
    his Goods for exhibition;
Every body ran him down
    and laughed at his position;
They thought him all the world behind
    —a goney, muff, or noodle.
Laugh on, good people—never mind
    —says quiet Yankee Doodle.

The Smithsonian catalog reveals an astonishing range of products and fluidity of free minds at work. It shows the divesity of forms our industry took, the wild variety of goods, the sense of pleasure in making a new world. It helps us see why, by the mid-eighteenth century, it was becoming England's turn to listen to the engines of change in America.

But one British industry remained hard to replicate in the New World. The English textile industry was by now powered by superior stationary steam engines. If we lagged behind England in building those engines, we still had higher hills and more rivers. We could do wonderful things with water power. In 1810 the New England trader Francis Cabot Lowell decided to create an American textile industry. Lowell went to England to spy on their equipment and came back to America to re-create the English textile technology on the Charles River in Waltham, Massachusetts.

The Waltham mills soon wanted more power than they could get from the Charles River. So the mill owners moved north of Boston to Pawtucket Falls on the Merrimack River. There they cut a Martian landscape of canals and millraces to supply the largest array of water-powered textile mills the world had ever seen. They built the new industrial city of Lowell.

By 1836 twenty textile mills in Lowell, powered by huge batteries of waterwheels, were producing fifty million yards of cloth a year. They

One of the Lowell mills, rebuilt and running today.

employed eight thousand people. In the 1840s the waterwheels began giving way to the new Francis water turbines, and the factories kept growing. The site was soon generating nine thousand horse-power—modest by today's standards, but a huge enterprise in those days.

Most of the workers were young immigrant women. They lived in tightly controlled boarding houses; no drinking or debauchery was allowed. The companies regulated their lives. They ran churches and sponsored cultural events. They boasted that these were utopian communities and invited anyone to compare them with English sweat-shops. A totalitarian air hovered over their utopia, of course; but what nineteenth-century industry was democratic? Promoters called Lowell the "Venice of America."[16]

As time passed, later generations of mill owners forgot about building workers' utopias. By the late nineteenth century Lowell had become just one more big, ugly industrial town. Yet Lowell had been a first step toward American industrialization. Historical preservationists have now rebuilt the old millraces and water-supply canals. Sunday visitors can now stroll through the old mill works and experience something of the Venice of America. They can look firsthand at the ingenuity, fore-sight, and perhaps demagoguery that got us started as an industrial nation.

8

Taking Flight

The recurring fantasies of my childhood were dreams of flight. I doubt I differed from other children in my imaginings, and in my childish way I seriously tried to achieve flight. I jumped from the garage roof into snowbanks. I scaled trees and cliffs. I swung on ropes. It's a good thing my mother never learned just how hard I worked at leaving the earth.

Sprained ankles and bruised ribs eventually convinced me that my body was earthbound even if my mind was not. I turned to model airplanes. I lived inside those lovely, light, buoyant structures. They carried me with them into the sky. My inner eye gazed down on the land from their vantage above.

This craving to fly is bred in the bone of our species. The old legends come out of the past with such conviction that we know some core of truth must undergird them. In Chapter 2 I refer to documented experiments with flight in the ninth and eleventh centuries. The Chinese flew humans in kites as early as the sixth century.[1]

One of the oldest and oddest intimations of early flight came out of the Cairo Museum in 1969. An Egyptian doctor named Khalil Messiha was studying the museum's collection of ancient bird models. He found that all the models but one were similar. That one was made of sycamore wood. It was a little thing with a seven-inch wingspan. It caught Messiha's attention because he saw it through the eyes of his childhood. He remembered the shapes and forms he had worked with when he built model airplanes as a boy. This was not a bird at all; it was a model airplane, and that was impossible.

Yet the other birds had legs; this had none. The other birds had painted feathers; this had none. The other birds had horizontal tail feathers like a real bird. Perhaps that was the most important difference. Birds do not have to be stable in flight because they can correct their direction; but a model airplane needs a vertical rudder to keep it moving straight. This strange wooden model tapered into a vertical rudder. One can also see that the wing has an airfoil cross-section. It was all aerodynamically correct. Too much about the model was beyond coincidence. Messiha's brother, a flight engineer, reproduced it in balsa wood and launched it. It flew!

This model had been dug up in Sakkara a hundred years ago. It was from the third century B.C., which meant it came out of the Hellenistic age of invention. Eighteen hundred years later, Leonardo da Vinci was still trying to invent flapping-wing airplanes and corkscrew-driven helicopters. But here an Egyptian had produced something with all the features of a modern sailplane.[2]

We have no evidence that anyone actually built a large version of this thing. Yet no one could have come this close to the real shape of flight without working on a larger scale. This little wooden model could hardly have existed unless someone had worked with large light models, or even with full-sized versions.

Archaeologists have never found a prototype. A large model, light enough to fly, would be too delicate to withstand the ravages of twenty-three hundred years. The original, if there ever was one, has long since joined the desert dust. Whatever form this Egyptian airplane might have taken, it has long since returned to the world of dreams and imagination from which it first came.

But the dream that arose in my childhood with such demanding and even life-threatening force is one that has marched through the ages since time immemorial. How else could those lines from the movie version of *The Wizard of Oz* be as compelling as they are? "Somewhere over the rainbow...birds fly.... Why then, oh why can't I?"

Flight touches a nerve. None of our technologies has obsessed us so long, and been so hard to convert to reality, as flight. None has presented such a vivid array of tales. None has so completely consumed inventors. In this chapter it is possible to dip only briefly into this huge arena of human endeavor and to sample a vast investment of human will and hope.

The myth of Daedalus and his son Icarus has stood through the millennia as the archetypal story of the dream of human flight. For thousands of years it has been the icon of our elemental craving to fly. Daedalus was a mythical Greek architect and sculptor. When he offended Minos, the king of Crete, Minos threw him and Icarus into prison. Daedalus made wings of wax and feathers. He and Icarus used the wings to fly to Sicily and to freedom. In some versions, Icarus flies too high—too close to the sun. The wax melts and he falls to his death. By now rockets have carried us completely free of Earth's gravity, yet no one has come close to duplicating Daedalus's flight under human power.[3]

In 1985 a team of engineers from MIT set a more modest objective, but a fearsome one nevertheless. They set out to fly a human-powered airplane from Crete not to Sicily, five hundred miles away, but to the island of Santorini, seventy-four miles north of Crete. Even that was over three times the existing world record for human-powered flight.

They built a wild dragonfly of an airplane with a pedal-driven propeller. It was made not of wax and feathers but of carbon-fiber composites and plastics. It gave a whole new meaning to the word *spindly*. Its wingspan outreached that of a Boeing 727, but it weighed only seventy pounds. Naturally they named it *Daedalus*.

The most serious problem was human endurance. The team carefully studied anatomy and metabolism. They did extensive testing of twenty-four men and one woman and finally gave the nod to a Greek bicycle champion, Kanellos Kanellopoulos. He would have to burn his body energy at the rate of one kilowatt for four hours running.

To sustain him on the trip, the team developed a special drink that would maintain balances of glucose, electrolytes, and water. Kanellopoulos would have to drink about a gallon of the stuff during the flight.

Armed with this witch's brew, he made the flight in April 1988. He flew the seventy-four miles in four hours to set a remarkable world record. Only one thing marred the success. As the plane approached the coast of Santorini a powerful crosswind caught it and snapped its tail boom. It splashed down safely, just thirty feet from the shore. This *Daedalus* seems to have carried just a bit of Icarus with it.

Somewhere over the rainbow we have always craved to fly. When we leave engines behind, we reclaim a whisper of the magic—the myth—that we so need to keep with us. And someday inventive minds will find a way to make Daedalus's flight all the way to Sicily, all the way from reality back to myth itself.

No wonder Joseph and Etienne Montgolfier's first untethered, manned, hot-air balloon flight on November 21, 1783, was another spark in the tinderbox of the western imagination. It was just four months earlier that sixteen-year-old John Quincy Adams and seventy-seven-year-old Ben Franklin had watched as Jacques-Alexandre-César Charles tested an unmanned hydrogen balloon in Paris.

For a while everyone was flying balloons and reaching altitudes limited only by their ability to breathe rarefied air. Jean-Pierre Blanchard first flew four months after the Montgolfiers; then he took his balloons on the road and became the first barnstormer. In England, expatriate American doctor John Jeffries hired Blanchard to fly him from England to France in a balloon. They had no common language, but that did not prevent them from achieving a fine mutual dislike as they skirmished over funding and credit for the flight. Still, they made the first aerial crossing of the English Channel on January 7, 1785. They barely made it. First, they had to dump everything loose overboard and finally they urinated their last few disposable ounces into the cold waters below.

Blanchard was an experimenter—first to drop animals in parachutes and first to try controlling his flights with sails and rudders. All this cost him more money than he could raise, so he took his act to America, where he hoped to do better. He arranged to make the first untethered American flight in Philadelphia on January 9, 1793. The Quakers had built a model prison with walls that offered means for hiding his takeoff from nonpaying observers. He cut a deal with the Quakers and then advertised in the *Federal Gazette*: Come watch the ascent for five dollars a person.

He collected $405 against $500 worth of expenses, and he took off before a crowd that included President Washington. When he landed in New Jersey, he served some remaining wine to local farmers who, in return, carted his balloon into town on their wagon.

Blanchard died sixteen years later after suffering a heart attack during his fifty-ninth flight. Afterward, his second wife, Marie, continued the act. The problem was, he had used hydrogen in preference to hot air in his balloons, and she made the tactical error of improving the show with aerial fireworks. Her flights were fine successes until 1819. Then she set the hydrogen on fire over the Tivoli Gardens in Paris. She managed to put out the fire, but only after losing so much hydrogen that the balloon fell to a rooftop. She might have survived, but the wind caught the empty balloon and dragged it over the edge. She died in the fall to the street below.[4]

That's a terrible ending, of course, but self-preservation was low on any early flier's list of priorities. Those primordial balloonists were driven by excitement. Great technologies are always born in passion and enthusiasm far more than in any pursuit of purpose.

England was, as we noted in Chapter 6, a much different world than Paris, and so too was her view of ballooning. Dickens began his *Tale of Two Cities* by calling up the country's differences, and by referring to the late eighteenth century as, "the age of wisdom…the age of foolishness…the epoch of belief… the epoch of incredulity." It was also an age of stimulus. Revolution was brewing in both countries. The English middle class applied religious zeal and technical creativity

Artist's image of the death of Madame Blanchard, from *Harper's New Monthly Magazine*, 1869.

to the improvement of life, while the French intelligentsia attacked tyranny with a highly honed and playful curiosity. So it is no surprise that the English looked down their noses at the new French mania for flying in balloons. An English newspaper called upon "all men to laugh this new folly out of practice as soon as possible." The first balloon flew in England as early as 1784, but it was not flown by an Englishman. The flyer was Vincent Lunardi, a dashing, self-aggrandizing young ladies' man from the Italian embassy. He made his own monument to the flight and described it thus:

> Let posterity know, and knowing be astonished!
> That on the 15th day of September, 1784,
> Vincent Lunardi of Lucca in Tuscany,
> the first aerial traveler in Britain,
> mounting from the Artillery Ground in London,
> and traversing the regions of the air for 2 hours and 15 minutes,
> in this spot revisited the earth.[5]

Lunardi barnstormed around England for two years, charming the public with his showmanship. Then, in 1786, his unoccupied balloon got away from him with a young bystander's arm entangled in one of its ropes. It carried the poor fellow a hundred or so feet into the air before he came loose and fell to his death. The English public and press promptly turned on Lunardi. A contemporary ballad ridiculed him:

Behold an Hero comely, tall and fair,
His only food phlogisticated air,
Now drooping roams about from town to Town
collecting pence t'inflate his poor balloon.

A beaten Lunardi returned to Italy and there took up ballooning with renewed panache. When he landed in a Spanish village, he was taken for a saint and triumphantly carried off to the local church.

The second English ascent was also made by a foreigner. It was by Blanchard, who, though he had all of Lunardi's megalomania, had none of his charm. Balloons were not born of eighteenth-century English virtues. Flight has always been the gift of less serious people, people driven by frivolous intellectual curiosity and risk. No wonder nineteenth-century America was soon trying to get into the act. Nothing makes that point quite so dramatically as does the story of Rufus Porter's dirigible.

Inventors had begun trying to create rigid navigable airships just after the first manned balloons. Henri Giffard finally capitalized on two generations of failure when he flew his three-horsepower, steam-driven dirigible over Paris in 1852.

Gold had been discovered in California in 1848, four years before Giffard's flight. The spring thaw of 1849 found everyone trying to get to the gold fields, but it was a daunting journey no matter how you went— two thousand miles over scarcely charted wilderness or an eighteen-thousand-mile ocean voyage around Cape Horn.

Just before gold was found, a man named Rufus Porter had flown some nice dirigible models. Now there was motivation for funding the real thing. Early in 1849 he published a pamphlet titled *Aerial Navigation: The Practicality of Traveling Pleasantly and Safely from New York to California in Three Days.*

He was serious. The pamphlet described plans to build an eight-hundred-foot, steam-powered dirigible with comfortable accommodations for a hundred passengers. It would go a hundred miles per hour. That was pretty grand thinking in 1849. Still, his specs were not far from those of the great zeppelins that flew eighty years later. Porter went on to advertise that New York–to–California service would begin in April. He wanted a fifty-dollar down payment on a two-hundred-dollar fare.

He began building immediately. His first *Aeroport*, as he called it, was actually only 240 feet long. But it was destroyed by a tornado. Later that year, he began a 700-foot version with new backers and more support. During a showing of the almost-complete dirigible on Thanksgiving Day, rowdy visitors tore the hydrogen bag. It could have been fixed, but rain got in and waterlogged the whole thing. So he started a third dirigible. A new round of technical troubles ended that effort in 1854.

Porter was oh so close to success. All he actually flew was a series of large steam-powered models, yet his ideas were sound, even if his dream was too large. It's as though the person who invented the boat had tried to begin with the *Queen Mary*. Still, his ideas about the internal structure of the dirigible, and partitioning the airbag, were eventually used in the successful airships. If Porter had been a better manager and money raiser, he could well have flown first. He honed the technologies that gave us the really grand zeppelins. The people who did succeed were buoyed by the very magnitude of Porter's eerie, visionary, gold-driven dream.

In fact, it was only a very few years later that a twenty-five-year-old German cavalry officer—a Prussian nobleman—came to America during the Civil War as a foreign observer of the Union Army. I'll withhold his name for a moment, because I don't want to give away the story. This young man had quite an adventure here. He narrowly escaped capture by the Confederate Army in Virginia; he watched draft riots in New York; he flirted with young ladies on a Great Lakes boat from Cleveland, Ohio, to Superior, Wisconsin; he ate muskrat and hunted with Indians. His remarkable journey eventually brought him to the International Hotel in St. Paul, Minnesota.

Just across the street from his hotel, a balloonist named John Steiner was flying passengers in his observation balloon. Steiner had flown for the Union Army as a civilian observer. His work probably saved McClellan's army from defeat in 1862. But he had quit the army in a dispute over pay and gone barnstorming.

The German officer decided to add a balloon ride to his American adventure. So Steiner sent him on a solo flight at the end of a seven-hundred-foot tether rope. Our young officer wrote an account of the experience. Outwardly, it was straightforward reporting of the military potential of observation balloons, but between the lines bubbled barely

The Zeppelin VII, *Deutschland,* as pictured in the 1911 *Encyclopaedia Britannica.*

controlled excitement. Back in Germany he finished a long military career. Then, four decades later, he found his way back to that bright moment of his youth. By the time he was sixty, many people (starting with Giffard) had built rigid navigable balloons, or *dirigibles,* but no one had yet made one that was commercially viable.

Our German officer's name was Count Ferdinand von Zeppelin, and he took up dirigible building just after his sixtieth birthday. His first airship flew in 1900. Zeppelin managed to synthesize the various elements other inventors had played with all through the nineteenth century. He lived and worked for another fourteen years, and by the time he died he had created the grandest machines in the air. The spectacular zeppelins continued to dazzle everyone until the whole technology went up in smoke with the *Hindenburg* crash in 1937.

It seems astounding that the seed for all this was sown in western America during the Civil War. Shortly before he died, Zeppelin wrote, "While I was above St. Paul I had my first idea of aerial navigation strongly impressed on me and it was there that the first idea of my Zeppelins came to me." Perhaps Zeppelin was so successful just because he was fulfilling a dream as atavistic as Daedelus'. It was a dream that had grown within him for a whole lifetime.

So we tried all down through the nineteenth century to fly. By 1903, when the Wright brothers flew, we had a highly honed technology of lighter-than-air flight. We had also spent a huge amount of human energy on heavier-than-air flight of two kinds: flapping-wing machines and helicopters. We never have mastered flapping-wing flight, but helicopters are another matter.

Leonardo da Vinci had the seductively simple idea of pulling himself into the air with a vertically mounted propeller. For the next four and a half centuries one inventor after another ran into terrible problems when he actually tried to do it.

By now, heavier-than-air flight has taken three basic forms. An airplane lifts off the ground when its propeller or jet pulls a lifting wing through the air. An autogyro also has a propeller that pulls it forward; but instead of a wing it has another large propeller on top. The second propeller is freewheeling—it is not powered. It just whirls in the wind and lifts the plane. Finally, the propeller of a helicopter is powered and it lifts the machine directly upward. It combines both power and lift in the same place.

After Leonardo, the idea of the helicopter resurfaced in France in 1784 in the form of a working model driven by a bowstring— about the same time that ballooning got its start there. During the nineteenth century, all kinds of ingenious helicopter models were built throughout Europe.

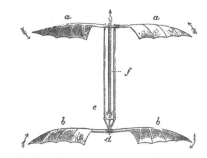

By 1874 the twenty-four-year-old French experimenter Alphonse Pénaud, had built a rubber-band-powered toy helicopter that he stabilized by equipping it with counterrotating propellers, one above, the other below. That toy model was sold to children

Penaud's toy helicopter as pictured in the 1897 *Encyclopaedia Britannica*.

through the nineteenth century. Finally Bishop Milton Wright brought one home to his boys Orville and Wilbur, and the rest is history.

And so attempts to build helicopters escalated. Enrico Forlanini flew a large steam-powered model to a height of over forty feet in Milan, Italy, in 1877, and in 1907 the Frenchman Paul Cornu hovered just off the ground for twenty seconds in a delicate double-propeller helicopter. Cornu, like the Wright Brothers four years earlier, was a bicycle maker. Other early helicopters were made, but they were all underpowered and hard to control. Then the more manageable autogyro was developed in the 1920s and attention shifted away from helicopters.

Igor Sikorsky built the first real helicopter in the United States in 1939. He had tried to build one in Russia thirty years before but failed. Then, after designing airplanes for thirty years, and using vastly improved technology, he succeeded. The Germans had built a successful hybrid helicopter-autogyro in 1936. In this case the engine drove both the lifting *and* the pulling propellers. About the same time Sikorsky built his first pure helicopter, the Germans dropped the forward propeller on their model, making it into a pure helicopter as well. The Russians soon copied the Germans, and military helicopters were serving both sides during the late years of World War II.

The helicopter was in people's minds long before the airplane. But it was a hard dream to fulfill. Its history is littered with half successes. The very simplicity of combining power and lift in one big propeller leads to terrible design problems. Leonardo was drawn in by its simplicity five hundred years ago, but he couldn't see how hard it would be to control motion with a single propeller. So complexity, masked as simplicity, kept right on teasing and misleading designers until the propeller-driven airplane had been a reality for thirty-six years.

Another feature of the new airplanes teased inventors. The problem of finding clear space for takeoffs and landings dogged airplane builders from the start. Several people tried to use water instead of land even before the Wright brothers, and the Wrights took off from a pair of rails. The first person who actually flew off water was Henri Fabre. In 1910 he flew his flimsy seaplane for a mile and a half in a harbor near Marseilles.

Three years later the first commercial flying service used an early seaplane. Twice a day it made the twenty-mile flight between Tampa and St. Petersburg in Florida. The plane could carry only one passenger, and the trip cost that person five dollars.

Twenty years later commercial seaplanes, or "flying boats," were the biggest things in the sky. It felt safe to cross the ocean in a flying boat, and that's how they were used. Pan American began service down to Central America with a ten-passenger Sikorsky seaplane in 1928. By 1934, these were being replaced with big flying boats such as the Martin M-130. It had a three-thousand-mile range and could carry forty people. Grandest of the successful flying boats was the Boeing 314, nicknamed the "Yankee Clipper." Pan American flew Yankee Clippers between 1941 and 1946. They had almost the wingspan of a 747, and they could carry seventy people more than four thousand miles.

Several perfectly enormous flying boats were built after the Yankee Clipper, but they came at the wrong time. Biggest of all was the Hughes Hercules—better known as the "Spruce Goose." Its wingspan was half again that of a 747, and it was designed to carry seven hundred people. In 1947 Howard Hughes flew it thirty feet into the air over Los Angeles Harbor. Then he put it back in its hangar, where it sits today—in mute and perplexing testimony to his convoluted thinking.

But Howard Hughes probably saw what was coming. When big propeller-driven airliners and then jets came into service after World War II, seaplanes lost their advantage. These land-based planes could fly the ocean, and there was no profit in using one kind of airplane over water and another over land. Furthermore, seaplanes were inherently large-bodied, high-winged machines, ill suited to near-sonic speeds. Of

Taking Flight

course, their large bodies made them wonderfully spacious. They typically had two- or three-story interiors. They looked not unlike whales.[6]

Small seaplanes are used today in Canada and Alaska, where lakes are easier to find than landing strips. But the big flying boats had a rather brief day in the sun. And the last of them represented a terrible misreading of the way the technology of flight was headed.

By now, we've all flown far over the rainbow. The experience loses its mystique when we are packed in rows, riding above clouds and without any intimacy with the wind. My father, who flew airplanes made of canvas and wood in World War I, told me about an experience in the early 1930s. He was riding on an early commercial flight in a modern passenger plane—a DC-3, I suppose, but I'm not sure. The pilot learned that he had a wartime flyer on board and offered my father a tour of the cockpit. "Here, take the controls," the pilot said. My father did. He made a large full-body motion—the kind of movement needed to make the controls of an early biplane respond—and inadvertently sent the aircraft veering off across the sky. Flying an airplane had been reduced to micromotions of the fingers. It was no longer something that involved the pilot totally.

For my father, flight had been made small. It had been turned from the lofty work of little boys jumping off barns into one more business for grown-ups. I doubt that flight will ever be quite what it was from 1783 until the debut of the modern passenger liner. But no matter. Other atavistic cravings remain, and technology will find new ways to transmute them for us, just as it has transmuted Daedalus.

9

Attitudes
and Technological Change

We look into the mirror of our machines, but what do we really see when we look in that mirror? How does change occur in the context of the mirror? The mirror turns out to be a strange reflector. We do not see *ourselves* when we first look at a new machine because there is a time lag in the reflection.

If you are a baby boomer or older, remember the first time you saw a computer. You felt neither need nor empathy for it. We cannot need what we have never experienced; yet that first glimpse initiated a long process. You have friends who still jitter about this new medium, wondering whether to accept the change it will bring into their lives or to keep dodging it. The need for transformation lies at our biological core, but we fear change nonetheless.

The first computers I ever used were so large that they filled rooms, and we had to speak to them with punched cards. The simplest conversation could stretch into weeks. We would submit three-inch decks of cards, wait twenty-four hours, and be handed a five-hundred-page sheaf of nonsense output because a do-loop went mad when we misplaced a period. During the 1960s we began to compute things that had been beyond us a few years before; but even as we did we grew desperately frustrated. All we talked about was increasing the speed of calculation, but what we really needed was a more accurate mirror of our human nature.

We finally began speaking directly to computers with keyboards during the 1970s. Then we realized we could compose text on the computer and print it out. Since the computer took no responsibility for organizing the text, we began to demand that word-processing logic be built into the computer. With the early 1980s, commercial software came on the market—canned sets of commands we could call up from the keyboard. Software now processed our words and laid our numbers out in spreadsheets. New programming languages removed more and more of the burden of speaking in the language of the machine. They became more fluent in human tongues.

While the computer grew more human, we also adapted ourselves to the computer. We changed our work habits and the character of our prose. We changed what we expect of human communication. Our concept of calculation has been altered completely as the computer has swallowed the old algorithms of multiplication and long division. Meanwhile, like another human being, the computer does more and more of its work behind our back.

How much thought did we give to the first photos we saw of IBM computers, isolated in "clean rooms" with their big tape drives? When my father saw his first automobile chuffing past an Illinois cornfield, he had no idea he would live his last days in a city completely reshaped by that primitive vehicle. Nor did he imagine how cars would eventually shape themselves to human bodies and human responses.

He understood these things no more in 1900 than I did in 1959 when another student in my research group told me he was using a computer to do a calculation. If he had told me he was changing human history, I would have laughed at him. But he was. For he had begun one of the very mirroring processes that has shaped the human species.

Processes much like this were going on long before the computer or the automobile. In this chapter I search out a few historical instances of the complex mirroring processes that define technology-driven evolution. We trace the way a few technologies and our attitudes toward them have changed each other. These examples offer just an inkling of how complex and hard to see these processes of change can be.

So let us go first to the eighth century A.D., to the near wake of the Roman Empire, when western Europe was still a primitive outback. Europe finally emerged as a new center of civilization because medieval engineers developed the water and wind power that the Romans (with their dependence on slaves) never fully exploited. But those developments had to wait until European agriculture became productive enough to support towns with masons and artisans—people free to create power sources. That, in turn, required a more powerful beast than the plodding ox to pull plows through the heavy, wet northern European soil. It required that the *horse* be integrated into European farming.

When, in the middle of the eighth century, Frankish kings began breeding large numbers of horses for military use, three things made them unsuitable for farming: (1) their hooves became soft and vulnerable to fracture in damp soil; (2) when they were harnessed in an ox yoke any heavy load would press upon their windpipes and strangle them; and (3) horses needed a better diet than oxen. They could not just graze grass but had to have a source of crude protein as well.

The nailed horseshoe and the horse collar solved two of these problems when they were introduced in the ninth century. The solution to the difficulty of feeding the horse was more complicated, but it also appeared in the ninth century. Until then, farmers had maintained two fields. They farmed one and kept the other fallow. This avoided the problem of robbing the soil of nutrients and leaving it unproductive. Then someone found that a field could be used two years out of three, if it was planted with one crop in the fall and a different crop in the spring a year and a half later.

Farmers could break their holdings into three fields. One would be planted with wheat or rye in the fall for human consumption. A second would be used in the spring to raise peas, beans, and lentils for human use and oats and barley for the horses. The third field would lie fallow. Each year the usage would be rotated among the three fields. We remember the spring planting in the nursery rhyme "Do you, do I, does anyone know, how oats, peas, beans, and barley grow?"

This clever scheme took two hundred years to adopt. The horseshoe and the horse collar were put to use directly, but three-field crop rotation required people to rearrange real estate and to change their social order.[1] For all its potential advantages, the three-field scheme was very hard to implement. The changeover was finally completed by the eleventh century. Only then could the great rebirth of European civilization follow. By the eleventh century, so small a thing as a horseshoe had so reflected and amplified through medieval minds as to help create a new civilization in Western Europe.

For contrast, consider another kind of case—an instance in which the changeover seems to have been almost immediate but was so invisible that historians are still trying to learn *when* it took place. The invention of the *mechanical clock* is hard to date, not because people wrote too little about early clocks, but because they wrote the wrong things. The mechanical clock seemed no more to the people who first used it than an improvement on the older water clock, which had been around for thousands of years. The records we have from those times dwell not on key working details but on outward appearance.

The water clock used a steady, regulated flow of water into a vertical tank. A rising water level in the tank indicated the time of day. That's simple enough, but water clocks were large, ornate structures with a great deal of supporting gear work and the same general fancification that was to be found on the mechanical clocks, which followed them. Water clocks often tolled the hours on bells and they usually had a face, just like the clocks today.

To a casual eye, mechanical clocks differed hardly at all from water clocks, but inside they used an escapement mechanism to regulate time (the balance wheel on a watch, or the pendulum on a grandfather's clock). As we saw in Chapter 5, an escapement ticks in a steady rhythm and lets the gears move forward at a steady rate in equal little jumps. The first escapement we know about was described in the mid-thirteenth century by the French architect Villard de Honnecourt. He did not exactly use it to control a clock. Instead, he created a charming gadget that steadily pointed at the sun while it moved through the daytime sky.

Monastery records for the next hundred years mention clock bells, gearing, and clock towers. But clock terminology rode right through the changeover. The first clear drawing of a mechanical clock was made by Jacopo di Dondi and his son in 1364. The di Dondis had probably been building clocks for at least twenty years by then. There is no way to be sure, but circumstantial evidence locates the invention of the mechanical clock even earlier, probably in the late 1200s.

It is strange to find that such an earth-shattering change could leave so few traces. Water clock inaccuracies had bottomed out at around fifteen minutes a day, and that's about as well as the first mechanical clocks did. Artisans working on mechanical clocks were now able to begin a rapid and sustained improvement of timekeeping accuracy. Between the fourteenth century and 1920 (when electrical elements were introduced into clockwork) clock accuracies doubled every thirty years. It was not long before those new mechanical clocks had swept the imagination of the Western world and led to new standards of precision—first in instruments and ultimately in thought itself.[2]

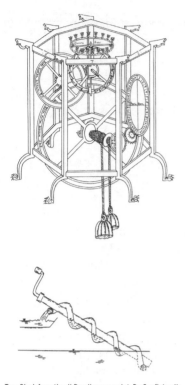

Top: Clock from the di Dondi manuscript *De Conficiendis Horologiis*, 1364.
Bottom: The screw pump of Archimedes, also known as the spiral of Archimedes.

The most important technology of an age is not always the most obvious one. More often than not, great change comes on little cat feet. It certainly did in this case, and it did in the matter of putting the horse to work on farms. The mechanical clock, which began the process of overturning Western thinking in the fourteenth century, had crept into the Western mind almost unnoticed during the thirteenth century.

Another complication in the adoption of any technology is the role that expectation plays in preparing its way. Take the story of Archimedes' screw pump. In the third century B.C. Archimedes invented a really clever pump, and it has been used all over the world ever since. It looks like a tube coiled around a long axle. You tilt the axle and put its lower end in water. Then you turn it. The open end of the tube picks up water and, as the coil turns, water passes from one loop to the next until it comes out at the upper end. The concept is subtle and hardly the kind of idea you would just stumble across.

Archimedean pumps were widespread in the classical world, and Roman authors described them. Well, they tried to describe them; but the configuration is next to impossible to describe in words alone. Archimedes' pump did not do so well during the high Middle Ages, when European attitudes were being shaped by the newly rediscovered philosophy of Aristotle. Aristotle had clearly separated motion into two kinds—motion in a straight line and rotary motion. Archimedean pumps combined both motions. They used rotation to move water upward along an axis, and that made them anti-Aristotelian.

But winds of change were blowing: In 1565 a Renaissance agricultural engineer named Giuseppe Ceredi patented an Archimedes pump. He systematically described the installation and use of batteries of these pumps for both irrigation and drainage. We might at first wonder how

Ceredi could be given a patent for a known device, but then we compare his dimensioned drawings, flow calculations, and economic analysis with the almost unreadable Roman descriptions.[3]

But even if Ceredi had found the idea in the old literature, he still put flesh and blood on it. After Ceredi, these pumps were quickly accepted across southern Europe. They were not, as one author put it, "something that would be created spontaneously by peasants." And they certainly were not something that people would take up naturally in a world that did not want to mix straight-line and rotary motion.

Not only did Ceredi have a right-brain ability to visualize and a left-brain ability to execute and organize detail; he was also able to break the straitjacket of Aristotelian thinking. A few years later Galileo took up direct combat with Aristotle's ideas about motion (as we have already seen in Chapter 5). Ceredi's reinvention of Archimedes' pump was a harbinger of that philosophical revolution.

It seems surprising to see a useful idea put on hold for so long, but the history of technology is filled with such cases. Take the English Channel. For thousands of years it has been a barrier that has tantalized people. That neck of cold, forbidding waters between England and France is at one point as little as twenty-one miles wide. We hear Shakespeare thinking aloud about Henry V's army as they prepare to cross the Channel to make war on France:

And thence to France shall we convey you safe,
And bring you back, charming the narrow seas,
to give you gentle pass;

In fact, crossing that treacherous neck of sea has seldom been a gentle pass. It has never been easy to charm those waters. So it is unsurprising that people have tried many means for getting themselves across. The first time the Channel yielded to anything other than a boat was two centuries ago, when Blanchard and Jeffries crossed it in their balloon.

The most primitive means for crossing the Channel, of course, is swimming. And we have no record of anyone trying to swim the Channel before 1872, a century after it was flown. The first person to succeed was Matthew Webb, who swam it three years later, in 1875. It took him twenty-two hours.

Like ballooning, heavier-than-air flight was attracted to the Channel almost immediately. In 1908 the London *Daily Mail* offered a £1,000 prize for the first Channel flight. That was only five years after the Wright brothers, and only two years after the first European flight. The French flyer Louis Bleriot won the prize a year later. Then, in 1979, a strange seventy-five-pound airplane called the *Gossamer Albatross* won the £100,000 Kremer Prize for the first human-powered airplane to fly the Channel. For three hours, pilot Bryan Allen pedaled its propeller and flew into the record books.

Only in the last decade have we managed the most down-to-earth means for crossing the Channel—that of tunneling under it. But that grand engineering accomplishment turns out to be remarkably old wine in a new bottle. The idea goes all the way back to the time of Napoleon. The English and French actually began a tunnel as long ago as 1881, but the British aborted it for fear it could serve the French as an invasion route. The British started another tunnel in the late 1970s, but they had to abandon *it* for lack of money.

England and France finally put fear of each other aside and completed their joint Channel Tunnel in 1994. To build the Chunnel, as it is called, they drilled two main train tunnels with a smaller access tunnel between them through the chalk marl beneath the water. French tunnels reached out from Calais to meet English tunnels from Folkestone, and one more work of technology finally overcame the formidable opposition that our attitudes had laid in its path.[4]

Now let us look at another kind of attitude problem: the failure to see the enormity of change while it is going on. The Reverend Dionysius Lardner wrote technical handbooks in the early nineteenth century. His first, published in 1828, was titled *Popular Lectures on the Steam Engine.* That was forty years after James Watt had brought the steam engine to maturity. Lardner's book rapidly went through a series of English and American editions. By 1836 it had been expanded to include everything from the availability of fuel to rules for railway investment speculators. But when Lardner speaks from the past about the power-producing potential of this new machine, we see vision stirred together with the same rash optimism we share with Lardner today. He begins by dramatizing the new state of affairs:

> In a [recent] report it was announced that a steam engine...erected...in Cornwall, had raised 125 millions of pounds, 1 foot high, with a bushel of coals....The great pyramid of Egypt

[weighs 13 billion] lbs. To construct it cost the labour of 100,000 men for 20 years. [Today it could] be raised...by the combustion of 479 tons of coals.

Then he goes on to calm fears about rising coal consumption:

The enormous consumption of coals in the arts and manufactures, and in steam navigation, has excited the fears of...exhaustion of our mines. These apprehensions, however, may be allayed by the assurance [of] the highest mining and geological authorities, that the coal fields of Northumberland and Durham alone are sufficient to supply [the present demand] for 1700 years, and...the great coal basin of South Wales will...supply the same demand for 2000 years longer.

Those reserves do little today to satisfy England's energy needs. But Lardner's failure to recognize a constant craving for more, and his sure faith that technological progress would prevent trouble in any event, are quite familiar to us today. His final assurance is one you've heard in discussions of population, energy, pollution, and every other modern problem:

[I]n speculations like these, the...progress of improvement and discovery ought not to be overlooked.... Philosophy already directs her finger at sources of inexhaustible power....We are on the eve of mechanical discoveries still greater than any which have yet appeared.[5]

Lardner certainly underestimated our appetites. But he correctly perceived the terrifying fact that human ingenuity will do more than we dare dream to meet frivolous wants as well as real needs.

And so optimism was the mood of the post–Industrial Revolution years. Optimism drove a perfectly astonishing technological leap forward. It was optimism that made us realize we could now use invention systematically to bend the world into whatever shape we wanted it to take. One person made that realization explicit. He was a young baron named Justis von Liebig. I first encountered Liebig's name back in 1953, when I was writing my master's thesis on the properties of a chemical called aniline.

I had just learned that aniline was a standard dye base that might make a good rocket fuel. Next I learned how to find data in a German periodical called *Liebig's Annals of Chemistry*. It was years later that I found out the remarkable way Liebig's life was interwoven with the aniline I was studying.

Liebig was born in Darmstadt in 1803. When he was twenty he went to Paris for a year to study with the famous French chemist Gay-Lussac. Gay-Lussac opened his eyes to the new idea that accurate experiments were needed to make sense of chemistry. Liebig came back to a post at the University of Giessen in Germany, and there he turned his young man's enthusiasm to the idea of precise experiments.

Liebig worked single-mindedly to set up a laboratory. He had to spend his own salary on equipment, but he was running a twenty-man chemistry research center by 1827. That may seem fairly pedestrian until we realize that nothing like Liebig's laboratory had ever existed before. We honor Liebig for work in organic, pharmacological, and agricultural chemistry, but that laboratory was his greatest contribution. Other chemists had to copy it to keep up. Before Liebig, research was an amateur's game. Soon these new laboratories were putting science in the hands of a new breed of professional.

In 1843 one of his former students sent Liebig an oil that he had isolated from coal tar. Liebig's lab found a compound in it that reacted with nitric acid to make brilliant blue, yellow, and scarlet coloring agents. It was a compound Liebig had already anticipated—a form of benzene with one hydrogen atom replaced by an amino group. They called it *aniline*. By 1860 Germany had built a new aniline-based dye industry that carried her into world leadership in industrial chemistry.

Of course, that leadership rested on Liebig's vision of systematic research, invention, and development. While the Germans were setting up their dye industry, Edison and others were setting up their own version of Liebig's laboratory here in the United States. You and I know full well what a profound impact R&D labs have had on American life. And it was all born of youthful energy. Our twentieth-century image of the scientific research establishment was born when a teacher, the great chemist Gay-Lussac, struck a chord in his twenty-year-old student Justus von Liebig.[6]

Whitehead would later call what Liebig had created "the invention of the method of inventions"—the beginning of our modern concept of the professional scientist and the professional engineer. No more dabbling with invention. Now institutions would hire these new professionals and go at invention systematically.

The nineteenth century was an epoch of change occuring too rapidly to be understood. Undigested change was everywhere. Fulton's steamboat, built when Liebig was four years old, is a good case in point. That steamboat was equipped with two sails. American riverboats quickly abandoned sail because rivers seldom gave them much room to navigate under sail. But abandoning sail at sea after using it for several millennia was quite another matter.

The American packet *Savannah* made the first transatlantic steamboat trip in 1819—that is, it got almost to Ireland before its coal ran out and it had to rely completely upon its sails. When the British *Great Western* established regular transatlantic passenger service in 1837, it still carried sail. So how long do you suppose it took to gain enough confidence to abandon the expensive backup protection of sails, masts, rigging, and extra crew?

The beginning of the end of sail traces back to the battle between the Yankee *Monitor* and the Confederate *Merrimac* in 1862. Those steam-powered, ironclad ships carried no sail because they too had been designed to serve as shoreline vessels. But the *Monitor* had an entirely new feature: In the center of the boat, where a mast might have been, there was instead a gun turret. By this time the conservative British admiralty was trying to replace the fixed guns on their ironclad warships with rotating turrets, but they found that masts and rigging interfered with the field of fire of a turret (a problem the *Monitor* did not have). The British clung to sail during the 1860s and built several ships with both turrets and masts. Those ships gave them a lot of trouble.

Not until 1871, sixty-four years after Fulton, did the English navy risk launching an oceangoing warship without any sail. It was the H.M.S. *Devastation,* and it set the pattern for future British sea power. But many merchant and passenger ships carried sail well into the 1900s—a full century after the first oceangoing steamboats.

We have to ask whether this was conservation of fuel or conservatism of mind. Some naval architects today talk about adding modern forms of sail to boost the power of merchant vessels. But though the engineers of the nineteenth century were many things, they were never conservationists. The long retention of sail represents an extreme instance of conservatism in engineering. Yet that conservatism is easier to understand if we remember the power of metaphor in the way we relate to technology. Sail was far more than functional technology when the first steam-driven ships appeared. Sail was woven into our language, our thinking, and our very being: "That really took the wind out of her sails"; "He was three sheets to the wind"; "Never weather beaten sail, more willing bent to shore"; "May the wind be ever at your back."

Marine artist FitzHugh Lane captured the situation when he painted New York Harbor in 1852. He recorded the busy harbor in a wealth of detail that early cameras might not have captured. Fifteen vessels were clearly identifiable. His foreground is dominated by two sail-driven packets, the new freighters that had been developed for economical runs between Europe and America. Further back is a conventional three-masted ocean ship. Scattered about are three small coastal sloops and three kinds of oar-driven boats.

By the time Lane painted his picture, maritime steam vessels were in wide use and we see them here. A third packet is driven by steam, and a riverboat is also in the harbor. The picture also shows two steam-driven towboats. One is a side-wheeler, which, like Fulton's first boat, is powered by an old Watt type of engine. The other, driven by a modern screw propeller, looks much like today's tugboats. Finally, in the background, you see one of the new ocean-going steamboats. It too resembles Fulton's first steamboat, with minimal sail and a pair of side paddle wheels.[7]

Now and then one of our technologies suddenly changes in form. That's what's happening in Lane's picture. Nine years before it was painted, my great-grandfather came from Switzerland on a sailing ship. He crossed the prairie to California on foot before the Gold Rush. Then, just after the Gold Rush, he was able to leave that new land on a steam packet to Panama. The change from sail to steam in the middle of the nineteenth century was that rapid.

Technological rollovers can be seen in photos of city streets taken sixty years later. Horse-drawn vehicles of every form move along with steam- and gasoline-driven autos and with a variety of bicycles. More recent photos of offices likewise show the last typewriters in a hopeless struggle for survival alongside the new word processors.

Isambard Kingdom Brunel's *Great Eastern*, was finished in 1858, and it carried a great deal of sail. Yet it was much more of the lineage of the *Titanic*. At their debuts, both the *Titanic* and the *Great Eastern* qualified as the largest ships ever made, and they are two of the best-known as well. The *Great Eastern* was almost seven hundred feet long. The

Titanic, launched fifty-three years later, was almost nine hundred feet long. And each suffered the same kind of accident soon after it was put to sea.

Isambard Kingdom Brunel, who designed the *Great Eastern*, was the greatest artist ever to work in iron. He was remarkably thorough, and the *Great Eastern* reflected that care. The ship was to be a passenger liner, and no cost was spared to make it safe. It had a double hull, and it was honeycombed with bulkheads that created almost fifty watertight compartments. But Brunel outreached himself with the *Great Eastern*. It was actually overdesigned and inefficient, yet it still provided transatlantic service for two years. Then, in 1862, it struck an uncharted rock in Long Island Sound and suffered an eighty-three-foot long, nine-foot wide gash in its hull. The inner hull held, and it safely steamed into New York Harbor.

The *Titanic* was another matter. When it was launched, transatlantic service had become a big, lucrative business. Bit by bit, safety standards had yielded to commercial pressures. The *Titanic*'s hull boasted a double bottom, but it had only a single wall on the sides.

It had fifteen sections that could be sealed off at the throw of a switch, but its bulkheads were riddled with access doors to improve passenger service. It had too few lifeboats, but the luxurious beauty of the ship was seductive. Why was it thought to be so safe? Historian Walter Lord says, "The appearance of safety was mistaken for safety itself."

When the *Titanic* gently grazed a North Atlantic iceberg in 1912, it suffered nothing like the continuous gash in the side of the *Great Eastern*. Rather, its plates were randomly punctured and sprung over a 250-foot length. But that was enough to put it under water within two hours and forty minutes.[8]

Image of the *Great Eastern* from a 1919 children's book. Note the masts and the large number of sails being used to augment the steam-powered paddle wheels.

And here we touch on an attitude issue that lies at the center of Chapter 15. Part of the blame for the *Titanic* accident was the very success of the *Great Eastern*. By 1912 so much success had bred a very relaxed attitude toward safety. We can see that pattern repeating itself throughout history. Before the loss of the space shuttle *Challenger*, for example, NASA's safety record had been unreasonably good. We forgot how intrinsically dangerous rocket launches can be.

So the *Titanic,* the *Challenger,* the Tacoma Narrows Bridge, and the *Hindenburg* all remind us that we engineers always have to mix a little fear in with our excitement when we design things. We have to remember that nothing so threatens our safety as early success.

Success takes other forms, however. Some technologies have matured into a truly enduring success. They have been burnished to a seeming dead end of goodness. We talked about the Indian canoe in Chapter 7. The modern violin uses the same basic technology that was perfected in Italy two and a half centuries ago. Most of the old Stradavarius, Amati, and other great violins have been fitted with new necks, bridges, and strings, and they are played differently today. But they are the same instruments that reached such a pinnacle of perfection 250 years ago that we can only try to replicate them today.

While early success is surely a danger, it is one that some technologies survive. The DC-3 passenger plane debuted in 1934. Viewed in the context of its function, the DC-3 was not improved by later airplanes. People who want small transport planes to get in and out of short landing strips and make short hops still buy DC-3s if they can get them. Today's airplanes seldom have to serve those functions, but when they do, the DC-3 is still a remarkably suitable airplane. Nobody will claim that later buildings improved on Chartres Cathedral or that word processors improved upon pens. Certain functions of the printed book have been taken over by the computer, but the printed book is nevertheless with us in perpetuity. Once a technology proves good enough to become woven into our psyche, it will survive indefinitely.

Much more might be said on the matter of attitudes and invention, but let us move for now to an attitude issue of a completely different kind. We turn in the next chapter to the technologies of ending human life—a business that has drawn in so many people that we cannot leave any broad discussion of technology without considering it.

Attitudes and Technological Change

10

War and Other Ways
to Kill People

We humans are a hardy lot. It eventually takes the cellular deterioration of old age to set most of us up for death, which then occurs by cancer, heart disease, pneumonia, or other illness. Death by natural causes is almost always the result of a protracted assault on our bodies. We are hard to kill.[1]

But now and then we undertake the technological problem of killing one another intentionally. That is seldom easy to do, and it has to play out against the universal human commandment "Thou shalt not kill." So the problem is not only a difficult one technologically, it is also one that calls up all manner of creative tactics of self-justification. The motivation for killing takes many forms—the greater good of society as expressed in war and capital punishment, mercy killing, personal gain (often expressed in crime against another person), revenge, anger, or suicide. I expect we all have sanctioned killing by one or more of these means at one time or another, by either words or deeds.

We have created little original technology for the purpose of killing one another. However, a great deal of our existing technology has been adapted to that purpose. Weapons for hunting have repeatedly been elaborated into weapons of crime or war. Lisa Meitner, whose 1939 paper described the energy release of nuclear fission, clearly thought she had identified the ultimate peacetime power source. Asked what use the Wrights' new airplane would be, Orville Wright unhesitatingly shot back, "Sport!" While war was far from the Wright brothers' minds in the process of invention, their first big commercial sale was to the United States Army.

The peculiar relation between creativity and killing comes home to me in my reaction to an event in the late days of World War II, when the war finally came closest to my quiet home in Minnesota. Since Tokyo was more than six thousand miles away, the mutual slaughter of Japanese and the Allies had largely been carried out in the Pacific Ocean.

Then in January 1945 we learned about Japan's secret weapon. She was trying to ignite our mainland with incendiary bombs. The bombs drifted over North America, carried by thirty-foot-diameter balloons made of fine Japanese mulberry paper called *washi*. The bombs were mounted on horizontal rings below the balloon, along with an array of small sandbags. The hydrogen expanded by day, and a balloon would rise until a sensor detected that it was too high and vent some hydrogen. As the balloon shrank by night, it fell until another sensor dropped sandbags. Finally, after three such cycles, the sensor lit a long fuse and vented the remaining hydrogen. The balloon then landed on American soil and, soon after, exploded.

We first thought that submarines off the West Coast were releasing the bombs. Then we realized the balloons were far too big and too numerous. The operation would have taken a huge number of submarines. So the U.S. Geological Survey set out to learn where the sand in the sandbags had come from. Sand carries a distinct fingerprint of mineral content, occasional coral fragments, and fossilized diatoms. (A diatom is the kind of single-celled creature that makes up algae.) That detective work showed that the sandbags had been loaded in a region near Tokyo.

Furthermore, we didn't know during World War II that Japanese meteorologists had discovered a jet stream, moving very fast at altitudes over thirty thousand feet. These bombs, it turns out, had made an astonishing journey all the way from Japan—something no one had thought possible. The balloons rode that jet stream and got here in a scant three days.

So bombs fell harmlessly near Klamath Falls, Oregon, and Bigelow, Kansas; they fell in remote corners of Manitoba, Colorado, Texas, and Mexico; they reached Iowa, North Dakota—even Michigan. None landed in St. Paul, but oh, Father Christmas, how this fourteen-year-old longed to see one! I gave no thought to death, only to fascinating new technology. Meanwhile, the Japanese press said that panic was gripping an America that now lay in flames.

To the best of our knowledge, only one balloon bomb actually killed anyone. That one landed in the Cascade Mountains. Five Sunday-school children and their minister's wife, on a fishing trip, found it just before it exploded, killing all six. But another balloon bomb caused mischief when it tangled in an electric line at the Hanford plant in Washington and cut the power. For a while it shut down the production of plutonium for the very atom bomb that would fall on Nagasaki only five months later.[2]

Complex as they both were, the atom bomb and the Japanese balloon bomb were both adaptations of peacetime technologies. Finding the means for killing people has not been much of a creative outlet for inventors. Instead, people who seek improved means for killing have forged a remarkable variety of swords from the plowshares made by creative people.

Of course we find circumstances in which the massive assault that is normally needed to kill a healthy body is no longer required. Such is the case when a person's life has been sustained beyond the reach of natural death in a hospital, or in an embryo whose life has claimed only a tentative beachhead. Killing calls for little technological help in these cases, and the ethical issues lie squarely before us.

But when a technological interface—a gun or a guillotine—lies between life and death, it can be much harder to come face to face with the ethical issues. In this chapter we shall see how technology clouds those issues by diverting attention from questions of life and death to the far more palatable question of inventing means by which killing is to be achieved.

If, therefore, my tone grows frivolous in this chapter, my intent does not. We engineers can make delicious puzzles of the problem of perfecting new killing machines. The intellectual fascination may last for a while, but when we look beyond our drawing board and see death, we had better have a very clear sense of what we are doing. In my lifetime I have been asked to look at problems connected with delivering napalm, poison gas, and even atom bombs. I have been asked to consult for cigarette makers. In every case I was first tempted by the complexities of the problem before I thought twice and walked away. We might face situations where killing is the lesser of two unavoidable evils, but that is the highest moral ground that killing can ever find. In the end, if we fail to value each other's lives (and the *quality* of those lives) above all else, we lose the right to call ourselves human.

One place where we all take part in ending lives is war, and the common wisdom says that war speeds up invention—that airplane performance, ship technology, and engine design all raced ahead during World Wars I and II—and that governments can speed the creation of ideas. The common wisdom accepts such assertions without a shred of supporting evidence.[3,4]

I shall use airplane speeds to explain my doubts, but any other technology would do as well. We all know how important it was to speed up airplanes during the world wars. Yet World War II airplanes such as the American B-17 bomber, the German Messerschmidt 109 fighter, and the British Spitfire fighter all existed before the war. The Spitfire, adapted from a peacetime racing plane, like most fighters at the start of the war flew about 350 miles per hour. By 1945 the advanced American fighters, the P-38s and P-47s, reached 420 MPH. The early German jet, the Messerschmidt 262, which was used in the waning days of the war, reached 585 MPH. But even it had been on the drawing boards before the war.

The remarkable fact is that throughout its history, the speed of flight has doubled every nine years. The rate of increase has been perfectly steady from the first primitive airships in the 1880s until orbital flight made speed a nonissue. The nine-year doubling was absolutely unaffected by war, depression, or presidential proclamations. The same story also holds throughout World War I. In 1914 the first scouting planes flew around 80 MPH. By the end of the war in 1918 the advanced SPADs could fly 134 MPH, and the increase is consistent with a simple doubling every nine years. In other words, once our creative energies were turned loose on the airplane, those energies went right on expressing themselves, war or no.

However, a government's wartime commitment to technology does increase production. And make no mistake, our miracles of production during World War II were dazzling. But human ingenuity is quite a different creature. It is remarkably impervious to external pressure.[3] We are told that necessity is the mother of invention, but history does not bear that out. The true mother of invention is a powerful, driving internal need to invent. We invent because we want to invent. The parents of invention are those inexorable internal needs for self-satisfaction and freedom.

It is easy to find case histories in which human ingenuity uses war as a vehicle rather than an objective. One of my favorites is from the dog

War and Other Ways to Kill People

days of the glorious high Middle Ages. By 1337 Europe had for over two centuries directed remarkable energy into the new technologies we talk about in Chapter 2—technologies that were both liberating and civilizing. But Europe had also begun diverting her energy into a set of eight Crusades. At first the Crusades reopened pilgrim travel to Jerusalem. They also opened up an east-west commerce of goods and ideas. Both Moslems and Europeans had the strength of religious tolerance and open-mindedness at the outset of the Crusades.

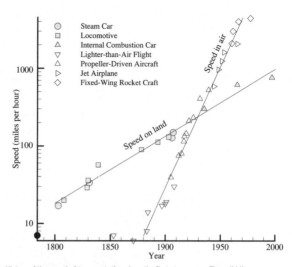

History of the speed of transportation since the first steam cars. The solid lines represent a pure exponential increase of speed with time. Notice that the data are uninfluenced by war or other major historical events.

The Crusades led both sides to trade that progress for the most self-destructive sort of prejudice and hatred. Fragmentary Crusades were still being waged during the Hundred Years' War (which began in 1337), but the Moslems finally drove the Europeans out of the Holy Land. After that, a dyspeptic Europe, engulfed in religious persecution and internal war, turned its bile on itself. A generation of bad weather, failed crops, famine, and susceptibility to disease had decimated the population of northern Europe. The plague would arrive ten years after the Hundred Years' War started, and it would eventually kill upward of half the European population.

Two years before the Hundred Years' War, a physician and engineer named Guido da Vigevano attached himself to Philip VI of France, whom he expected to go on an obligatory Crusade. To strengthen his position with Philip, Guido wrote a sort of Crusade handbook for him. Nine folios of the book advise the king on how to look after his health on the journey, and the other fourteen folios advise him in military technology.

Historian Rupert Hall points out muddy inconsistencies between Guido's text and his sketches of military apparatus. But Guido's devices are clear enough in their broad intent, and they represent a last breath of the soaring medieval imagination. Guido recognized that wood would be hard to find in the Holy Land, so he advised breaking siege equipment (which was usually built on the spot) into prefabricated elements that could be carried on horses. He said a lot about joints and assembly. He included folding attack boats and pontoon bridges. He designed two self-propelled battle wagons. One was crank-driven, and the other was powered by a sophisticated windmill. He proposed innovative body armor and siege equipment.[5]

King Philip never got to the Holy Land, and no one ever tried to build Guido's wonderful machinery. Two years after Guido presented his book, it was King Philip who *started* the Hundred Years' War by seizing an English-held duchy in what is now southwest France. Today, I look at Guido's marvelous Picasso-like sketches, without perspective or three-dimensionality, ideas tumbling one over the other, and I ask what it is I am really seeing. Guido gave us a kind of fantasy armory for beating back a fantasy enemy—a child's view of war—while the practical world around him was bent on setting loose straightforward technologies of destruction.

Among Guido's many counterparts in the long history of wars was Anthony Fokker, perhaps the greatest builder of military aircraft in World War I. Fokker clearly danced to the same drum; however, *his* killing machines, unlike Guido's, were completely effective.

The first airplanes that took to the sky in World War I had only one purpose: scouting enemy positions and movements. Still, in no time at all, their pilots looked for ways to shoot each other down. Fliers made the first kills by firing pistols and rifles off to the side. Then backseat observers began operating movable machine guns. But what they clearly needed were forward-firing guns that could bring an enemy down from behind.

The British were first to mount forward-firing guns on the upper wing so they could shoot *over* the propeller. But that made aiming hard, and it put the guns out of the pilot's easy reach when they jammed. The French took the next step. They put metal deflectors on the propeller so the pilot could fire straight through the blades, with a bullet glancing off now and then. That worked until crankshafts deformed under the hammering of the pilots' own bullets.

And here Holland's Anthony Fokker enters the scene. The year before World War I began, Fokker was only twenty-three and building airplanes.

Germany contracted with him to build ten airplanes, and he went to work. War broke out months later, and Fokker was suddenly Germany's man of the hour. By 1915 his monoplane, the Eindecker, was doing frontline scout work. Then the Germans brought him a captured French plane with metal plates on the propeller. Could he do that with the Eindecker? According to his autobiography, Fokker tells what happened next.

Fokker thought about the problem of firing through the propeller and realized that the solution was to let the propeller fire the gun. The propeller turns at 1,200 rpm, and the gun fires 600 times a minute. Fokker proposed putting a cam on the shaft and letting it fire the gun every other turn. Then no bullet would ever hit the propeller. Three days after the Germans brought him the French plane, Fokker came back with a synchronized machine gun.

The device worked well enough in tests, but German officers wanted a combat demonstration. They wrapped Fokker, a Dutch civilian, in a German uniform and hustled him off to the front. He took off in his Eindecker and soon spotted a two-seater French scout below him. He put the plane into an attack dive, located the scout in his sights, then realized he was about to kill two people! Fokker felt sick to his stomach and flew back to the aerodrome without firing a shot. Let the Germans do their own killing, he vowed. So the army relented. They sent a pilot named Ostwald Boelke up to try it out. Boelke went on to become Germany's first ace. Fokker went back to making the advanced German airplanes that killed thousands of Allied pilots throughout the war.[6]

The Allies called his mechanism the "Fokker scourge." After the war, we used Fokker's *commercial* airplanes here in America. My father, a newspaper writer, had been a pilot in France during World War I. He met Fokker in the United States. Later he told me what a thoroughly pleasant fellow he'd found Fokker to be.

The First World War provides a virtual laboratory for looking at the tenuous relation between creativity and killing. Barbara Tuchman called her book about World War I *The Guns of August* because the form of the war was set during its first few days in August, 1914.[7] The German attack plan, from a nineteenth-century military strategy book, promised victory in a few weeks. Yet the war turned into four years of human attrition along a double line of trenches that ran from Nancy west to near Paris, and then north to Ostend on the English Channel. By the time the carnage ended, eight and a half million soldiers had died—most of them in that 350-mile row of trenches.

Generals on both sides tried to fight the war the way Napoleon might have. They had plenty of warning that new technologies of slaughter would create a stalemate, but the warnings were misread. Napoleonic muskets had given way to bolt-action rifles such as the British Lee-Enfield. The machine gun had matured. These weapons laid down a field of fire that made an open-field attack impossible. Once troops were entrenched, they were immovable.

America's Civil War had already made it clear that even the first breech-loading rifles would make a mess of traditional warfare. If European generals were not about to take America's military advice, they could have paid more attention to their own Battle of Duppel in 1862. Six thousand ill-equipped Danes built a continuous fortified trench one kilometer long and held off eighteen thousand well-equipped Prussians. The Prussians went at the Danes as if they were attacking a conventional masonry fort.

Instead of crumbling under artillery fire, the protective dirt just flew in the air and fell back down. The Prussians finally overran the Danes, but it took two months and far too many lives. Then, instead of realizing that the Danish trench was a remarkable new defensive strategy, the Prussians simply thought conventional tactics had taken a bit longer than they should have.

Storm warnings like this kept coming. Open-field formations were cut to pieces in the Prussian-Austrian conflict in 1866, and again when the Russians fought the Turks in 1877. The only protection was earth. A year after the Danes fought off the Germans at Duppel, Union forces on the other side of the Atlantic were trying to retake Fort Sumter from the Confederates. Their long-range rifled cannon pounded the fort into rubble. But once Sumter was destroyed, Confederate troops simply burrowed into the rubble, from which they could not be dislodged. The Union Army had unwittingly given them their protective trench.[8]

So World War I settled into a double row of trenches where five thousand soldiers were killed in a slack week. A half million might be lost in a major assault. Meanwhile, the battle line hardly moved. The stalemate was finally broken in 1918 when the Allies came out with a new weapon, the armored tank, and a wholly new set of tactics for using it.

France was on the winning side, but twenty-two years later, in 1940, the French had built a kind of supertrench called the Maginot Line. This time Germany, with her modern Panzer tactics, simply punched a hole in it and took France within a week. With that victory, this terrible cycle in the technology of killing had begun yet another round.

And so the Second World War ground on, killing even more people than the first one had. Late in that terrible conflict—on February 17, 1944—American carrier-based bombers ambushed a large Japanese fleet in Truk Lagoon. Truk is a circle of coral, forty miles across, with eleven small islands in it.
It is a perfect natural anchorage in the western Pacific Ocean. Our bombs put sixty ships and thousands of Japanese sailors to rest in a stunningly beautiful cemetery. There they lie among coral, plumed hydroids, and white sponges, with curious damselfish and squids nosing through the tanks, skulls, and fine tableware that have found a strangely natural place in the fantastic landscape on the shallow ocean floor.

Here sits a three-man tank, stridently green in a cloak of marine flora. There are the controls for the stern gun of the *Fujikawa Maru*, festooned with orange sponges and soft corals. Below is a bin of huge spheres, blanketed in green and orange—live mines with small blue fish darting among them.

The Truk site adds practical and moral dilemmas to the perennial question of historical preservation. I do not speak metaphorically when I call it a cemetery. After the war, Japanese divers retrieved what remains they could find. But so many washed bones lie just beyond human reach.

The government of Truk protects this phantom navy from looters, but the ghosts also mount their own guard. In the waning days of the war, the Japanese ran low on the materials normally used to make explosives. They resorted to untried chemical alternatives in their ordnance. Old munitions are dangerous enough, but many of the bombs and mines in these hulks are so unstable that they have been known to go off spontaneously. It is not just that they cannot be moved; these aging agents killing cannot even be jiggled.

Oceanographer Sylvia Earle has swum through this landscape and brings up the issue of how it should be treated. The ocean will eventually eat it up, and, she argues, we should yield to that due process. Oil tanks will corrode and spill their limited contents. Munitions will gradually leak into the surrounding water. The small creatures of the sea will slowly weave the devastation of Truk Lagoon into the reefs around it. But for a while longer, brilliantly colored fish will keep steering divers through this remnant of World War II.[9] That terrible conflict will show itself to a few people in terms that no record book reveals. History will live, and education will be completed. And the next world war will not be started by people who still have this kind of intimacy with the last one.

The war at sea brings us back to another remark I made at the outset: that while war does not drive creative invention, it certainly does drive production. Nowhere has that been made clearer than in the instance of the *Liberty ship*. The United States had built only two freighters between 1922 and 1937. Our merchant-ship building was nearly dead on the eve of World War II, and the Axis nations were torpedoing Allied ships off the surface of the ocean. England, in particular, needed ships right away. To make matters worse, Allied shipbuilders were hopelessly preoccupied with warships; somebody else would have to start making freighters from scratch.

In 1940 the English were desperate enough to turn to an American group of heavy construction companies led by Henry Kaiser, who had never built any sort of ship. The English carried with them the plans for a kind of generic freighter. It was a simple, functional ship of World War I vintage. Kaiser had neither workers nor shipyards with which to make these ships. But he turned his lack of preparation to remarkable advantage. Did it take years to train a well-rounded shipbuilder? Kaiser rearranged work so he did not need well-rounded people. He broke shipbuilding into components and prefabrication so that each worker had to learn only a small piece of the job. Did he need heavy equipment to cut metal plate? No matter: He simply used oxyacetylene torches. In one case, he cut the time it took to train novices to tightrope across steel structures by hiring ballet dancers as fitters.

Kaiser redefined shipbuilding to match his resources. For the first time, he did it with assembly-line techniques—interchangeable parts on a gigantic scale. His product, the Liberty ship, was 440 feet long, and it carried nine thousand tons of cargo. The first one came off the ways just after Pearl Harbor. During 1942 ships were launched within less than a month after construction began, then in just ten days. Finally, one was launched after just four days' time. Kaiser ate steel so rapidly that he had to set up his own mill.

Behind all this schoolboy excitement lay a darker side. We produced eleven million tons of shipping in 1942, but we lost twelve million tons to submarines. In 1943 we raised that to twenty million tons of shipping, and we prevailed. The Liberty ship saved us.[10]

Kaiser's genius lay in the way he could hold shipbuilding up to the clear light of his own amateur vision and bring it into the twentieth century. What he accomplished was rooted in a powerful common purpose, which ended when the war ended. The ships themselves represented no creative impulse; they were anachronisms. But Kaiser had, without question, pulled off a production miracle when he actually built a ship, keel to launch, in only four days.

Proponents of the notion that war drives creativity inevitably talk about one invention in particular: radar. Yet the concept of radar is almost as old as radio itself. Radio pioneers Marconi and Tesla both pointed out that we could locate metal objects by bouncing radio signals off them, and as early as 1904 a German engineer named Hülsmeyer patented a radio echo device for locating ships at sea.

During the 1930s all the major powers worked to develop workable airplane and ship spotting systems that used radio waves. By the way, the acronym *radar*, which stands for "*radio detection and ranging*," was coined in 1942 when the United States Navy started using it. American army and navy engineers discovered in 1936 that they could detect aircraft at distances of more than a hundred miles when they used long enough wavelengths. They had mobile detection units in production by 1940. The first of these units were field-tested in Panama, and late in 1941 five of them were being field-tested in Hawaii.

One of the Hawaiian units was stationed on the northern tip of Oahu on the night of December 6, 1941. Private Joseph Lockard was training Private George Elliot, and they were to go off duty at 7:00 A.M., when a truck was to pick them up for breakfast. The truck was a little late, and Lockard was trying to give Elliot some extra time on the unit. At 7:02, Elliot saw a very large reflection, 136 miles due north of their position.

They tracked the signal for eighteen minutes; then Lockard called the information center, where the lieutenant on duty dismissed it as, in his words, "nothing unusual." Lockard and Elliot went on tracking the signal until 7:39 A.M., when the 183 Japanese dive-bombers and fighters that were creating it were only twenty miles away. Then the truck arrived to take them to breakfast, so they folded up their equipment and left. Sixteen minutes later the planes hit Pearl Harbor.[11]

By ignoring the signal, we lost three thousand men, dozens of large ships, and 80 percent of the airplanes on Oahu. Still, it is too easy to criticize shortsightedness. Radar was a new invention, and invention is alien or it would not be invention. We have to be introduced to new technology—gradually brought to understand what it can do. Unless it appears just when we are ready for it, invention must be championed. The great inventions that have revolutionized the world have seldom been recognized in their first incarnations. The lightbulb, the steamboat, and the telegraph had all been invented long before Edison, Fulton, and Morse came on the scene to show us their full potential.

And, for the United States, World War II began when we saw Japanese airplanes coming and did not believe our eyes. Pearl Harbor had to go up in flames before we learned to take radar seriously.

So much for the anonymous killing of war. Now let us turn to the killing of a specific individual by capital punishment, and let us begin with the *guillotine*. The history of beheading is all mixed up in class distinctions. In ancient Greece, Xenophon singled it out as a noble punishment. The Romans, who did horrible things to common criminals, also saved decapitation for nobler folk. They called it *capitis amputatio.*

William the Conqueror brought beheading to England, where it was also set aside for nobility—people such as Lady Jane Grey and Anne Boleyn. When the English beheaded the lower classes, it was only to finish off victims who had first been tormented in ways too nasty to talk about here. The reason for mechanizing such a seldom-used punishment was that axmen were often inaccurate. Victims, after all, paid executioners a gold coin so they would cut cleanly. A few early beheading machines were tried out. The sixteenth-century Scots used a device coyly named "the maiden," and an English machine called the "halifax gibbet" saw some use.

But it took the egalitarian French Revolution to bring beheading to commoners. Joseph Guillotin was a physician and a member of the Constituent Assembly in the early days of the French Revolution. In 1789 he got a law passed requiring that beheading machines be made so that the privilege of decapitation would no longer be confined to nobles and the process of execution would be as painless as possible.

The machine was built, tested extensively on dead bodies, and turned loose on common criminals in 1792. Of course, once this was done it became all too easy to dispose of counter-revolutionaries, and the slaughter called the Reign of Terror followed.[12]

The American adventurer and inventor Count Rumford (Chapter 3) gave an interesting footnote to Guillotin's invention. Rumford married the widow of the famous chemist Lavoisier, who had been among the thousands who died on guillotines. But a few years before his marriage Rumford wrote, "I made the acquaintance of Monsieur Guillotin the contriver of the two famous guillotines. He is a physician, and a very mild, polite humane man."

This may all seem quite ghoulish, but the point is clear enough. It is that we technologists are obliged to think twice when we are given the chance to sanitize death. When all is said and done, very little indeed separates gentle Dr. Guillotin's beheading machine from, say, the development of the neutron bomb.

Let us finish with a more contemporary instrument of capital punishment, the *electric chair*. Its story begins in 1875, when the inventor Nikola Tesla was a student at the Austrian Polytechnic Institute. Tesla brashly suggested that electric motors would run better on alternating current, and his professor asked where a bright student came up with such claptrap; there was no way to make an alternating-current motor. Six years later, in Budapest, Tesla walked through the park at sundown, reciting sad lines from Goethe's *Faust*. The aging Faust, who had failed to uncover the secrets of nature, thought about sunset and the end of life:

> The glow retreats, done in the day of toil;
> It yonder hastes, new fields of life exploring;
> Ah, that no wing can lift me from the soil,
> Upon its track to follow, follow soaring!

Then it hit Tesla: Maybe Faust was stuck, but *he* was not. He suddenly saw how the leading and following magnetic fields in motors and generators could be arranged to create and make use of alternating current.

Three years later Tesla went to work for Edison in the United States. He tried to interest Edison in alternating current (AC) but was told that the idea was downright un-American. Tesla and Edison soon parted company. Tesla managed to get funding from the financier J. P. Morgan, and he was issued a series of AC patents starting in 1887. He soon convinced George Westinghouse to put his money into the development of alternating-current power systems.

Edison's response bordered on maniacal. He launched an appalling campaign to discredit both Westinghouse and Tesla. The idea was to show that alternating current was too dangerous to use. He invited reporters to demonstrations where stray dogs and cats were placed on metal sheets and electrocuted with a thousand volts of alternating current.

Next Edison took out a commercial license to use alternating current. The world found out why after he had made clandestine visits to Sing Sing prison. He had created the electric chair. Now the American public would see what damage alternating current could inflict on a human being. Before the chair was first used on a fellow named William Kemmler, Edison's people started killing larger animals in their demonstrations. They asked, "Is this what your wife should be cooking with?"

When poor Kemmler was taken to the chair to be (as the Edison people put it) Westinghoused, the voltage was too low. A half-dead Kemmler had to be electrocuted a second time to finish him off. All this served Edison's purpose, of course. There was no need for Kemmler's passing to be a pleasant one.[13]

In the end, the widespread use of alternating current prevailed, but it took twenty years for Edison to admit defeat. The electric chair also persisted into our lifetimes. And I am inclined to think that it grants us a rare glimpse of that dark place in the collective human psyche where the idea of capital punishment is born. Human ingenuity does, no doubt, turn now and again to the task of ending human life. But I believe it is safe to say that when it does, it can hardly help but open the door to misuse.

11

Major Landmarks

oward the end of the twentieth century we saw countless inventories of inventions and achievements—the top twenty scientific breakthroughs of the millenium, the fifty most important people of the century, and so on. I am of two minds when I look at such lists. On the one hand, they rightly celebrate and draw our attention to much good that has been done. On the other hand, as we noted in Chapter 4, a vast portion of real accomplishment goes on below our level of perception. That point recurs in Chapter 14. So much of the creativity that defines us as people is inevitably left off such lists. It would be a futile exercise to correctly identify any definitive list of the most influential machines of all time.

Yet some inventions really are landmarks. I have selected from among the first year's *Engines of Our Ingenuity* radio programs a set of starting points that strongly propagate forward in time. Some might be obvious choices; others might not. But each can rightly be called a landmark because it sits squarely on some major highway of subsequent development.

There is, of course, only one place to begin such a list, and that is with the great progenitor of machines—the first machine most people will name. We begin with the *wheel*, which has become such a universal and familiar icon of our technological world that we forget the enormous conceptual leap it embodied.

The wheel was almost surely invented somewhere within the borders of present-day Iran or Iraq, five and a half millennia ago. That in itself is surprising because it happened so late in human history. The wheel was also confined within Europe and Asia for a long time. Wheels were hardly seen in the American hemisphere until European settlers began bringing them into regular use in the sixteenth century. There is evidence that the eleventh-century inhabitants of what we now call Mexico had the concept, but we have no evidence of its general use.

Of course, you and I have lived our entire lives with a hundred thousand different forms of the wheel. That experience limits our ability to imagine what a difficult concept it was to originate. Try to look at the wheel from the standpoint of someone who has never seen one, difficult as that is to do: You understand movement in a straight line and the idea of turning things around. But can you make a connection between the two? Can you conceive of making a vehicle go forward by turning something around? We have all played the children's game of patting our head and rubbing our stomach at the same time. It is very hard to do because it is difficult to conceptualize these two very different motions simultaneously.

The author first confronts the conceptual riddle of the wheel.

Here is a question to weigh as you consider wheels: About what point does a wagon wheel rotate? The obvious answer is the axle. That may be valid for an observer riding on the wagon. But the inventor stands upon static earth. For the earthbound observer, the wheel rotates not around the axle at all, but around the point where the wheel touches the ground. At that point, the moving wheel is stationary for an instant, while the axle moves at the speed of the wagon and the top of the wheel moves exactly twice as fast as the wagon.

The conceptual difficulties of the wheel are compounded if we move to a variation of the idea, the hand crank. Like the wheel, the crank is such a common device that we might think it had been with us since the dawn of history, but it has not. The hand crank has been in general use for only one millennium. The Greeks didn't have it. The Egyptians didn't have it. The vaunted Romans, with all their highly touted technology, never arrived at this seemingly elementary device.[1]

The hand crank, of course, takes the problem of converting the back-and-forth motion of our upper arm into a rotational motion, and it freezes this transformation into one location. In a sense, it requires that we solve the problem of patting our head and rubbing our stomach at the same time. The hand crank appeared in western Europe only in the

Major Landmarks

ninth century A.D., and even after that it was slow to find its way into general use.[2]

Historian V. Gordon Childe offers a helpful slant on the invention of the wheel. What we should be looking at, he says, is not the wheel itself but the use of rotary motion. The human wrist/arm configuration allows just about 360 degrees of rotation. That's why hand drills using back-and-forth hand motions are among the oldest tools. Late–Stone Age artisans extended that rotation with various pivoted devices. A primitive spindle plays out wool or flax

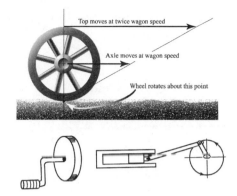

Top: The kinematics of the wheel. Bottom left: The simple crank (ca. AD 850). Bottom right: The slider-crank mechanism (eighteenth century).

fibers while the operator keeps it moving with thumb and forefinger. Early doors, with a vertical shaft on one side, were anchored in sockets that let them swing open and shut.

The next jump in sophistication was using bowstrings to drive the back-and-forth rotary motion of drills and fire starters. These devices all ran one way to a limit; then they had to unwind. *Continuous* rotation was the conceptual hurdle, and the two primary examples, the vehicle wheel and the potter's wheel, arose about the same time.

A potter's wheel is a horizontal turntable that holds a lump of clay and turns at least 100 rpm. Childe finds one potter's wheel from the region of the Tigris and Euphrates Rivers from as early as 3300 B.C. The earliest vehicle wheel turns up in a picture from same region in 3500 B.C. Those dates are close to each other, and examples are too rare to fix dates accurately. So, to the best of our knowledge, not just the wheel but continuous rotation itself dates from fifty-five hundred years ago in the Fertile Crescent. Another invention, closely kin to the wheel, was the simple *compass* that we use for making circles. The first hinged compasses can also be traced to that same region and the same period.

Early wheels show a progression of understanding. The first wheels were cut from large wooden slabs, built up of boards with the grain of the wood running across the wheel. Not only is that a clumsy design, but its aesthetics are disturbing as well. The grain of the wood seems to contradict rotational motion.

Not until 2000 B.C. do we find wheels with spokes. The spoke eventually introduced yet another counter-intuitive subtlety as designers realized they could make very light wheels by hanging vehicles upon the upper spokes. Since spindly spokes, such as those on a bicycle, carry no load at all in compression, only the upper ones serve any purpose in a given instant.

Other questions of rotation had to be answered as well. Wheels are best left free to rotate on a fixed axis. If they are anchored to a rotating axle, then they cannot turn at different speeds going around a corner. The idea of a swiveled front axle that can turn into a curve is barely two thousand years old. And so it is not the wheel itself but the problem of rotation that has kept dogging our minds. The ancient Sumerians recognized the problem. And we have (if I may) *spun* out the subtle ramifications ever since.

Most important ancient inventions seem to have been made over and over at different times and in different places. Not so the wheel. The anthropological evidence strongly suggests that it was a concept that originated in one place and then diffused (mutating and taking new forms) to other peoples and other cultures. The wheel, the solution to the problem of continuous rotary motion, was very likely the result of an *isolated* act of ingenuity.

The matter of rotary motion leads to yet another technological landmark, the *windmill*. Since windmills combine the kinematic complexity of wheels with the mystical attributes of the wind (Chapter 2), it is little surprise that Miguel de Cervantes made so much of them in *Don Quixote*. Quixote dwelt in the twilight of the age of chivalry in the fourteenth or fifteenth century but he was a creature of Cervantes, who lived in the late sixteenth century. Early in the story, Quixote cries, "Look there, my friend Sancho Panza, where thirty or more monstrous giants present themselves, all of whom I mean to engage in battle and slay!"

What he sees, of course, is an array of the power-generating windmills that dot the Spanish landscape. Windmills had rapidly come into wide use in Europe beginning two hundred years before Quixote and four hundred years before Cervantes. Waterwheels had been in wide use for centuries by then, but the new windmills were more complicated. They were at the mercy of the sometimes-fickle winds, but they delivered more power than waterwheels, and they made it possible to grind grain where there were no streams—the Dutch lowlands and the Spanish plains.

By 1760, windmills reached an astonishing level of sophistication. They were equipped with automatic regulators that controlled the speed of rotation, adjusted the pitch of the fan blades for maximum power at a given wind speed, and oriented the fan so it always faced directly into the wind. When windmills were used to mill grain, they were equipped with devices that regulated the pressure of the millstones on the grain. But it was also in the 1760s that Watt began developing a vastly improved steam engine. As the eighteenth century ended, windmill development was abandoned in favor of these new engines. Watt was the Quixote who really slew the windmill.[3]

Of course windmills did not go away completely. Today they are still a choice power supply for isolated use where there is no commercial electricity—especially for filling cattle watering troughs on the prairie. In the nineteenth century they provided the only means for pumping the water needed to resupply the railway-engine boilers. They played a large role in opening the American West. Latter-day engineers are now concocting a dizzying array of improvements in the hope of using windmills for electric power generation. The modern propeller-bladed windmill is three or four times more efficient than the advanced eighteenth-century mills, and forty times as powerful.

At the heart of any windmill is an extremely complex variation on the theme of continuous rotation: the *propeller*. Not only does the propeller turn like a wheel, but it must also be formed to make best use of the wind passing through it. The variable-pitch sails used in those mills over two hundred years ago were an innovation that airplane designers did not rediscover and apply to airplane propellers until the 1930s.

And that brings us to another major landmark, the *airplane*. We talk specifically about airplanes in Chapter 8 and that includes the Wright brothers' flying machine, which would be a landmark in anyone's book. But another airplane is a landmark in itself that deserves special attention. It is the Douglas DC-3, which we already acknowledged in Chapter 9.

I am now an old man, yet just the other day on a trip in a jet plane I noticed a small air service with several operational DC-3s parked outside its hangers. The import of that becomes clearer when I recall one of the first movies I ever saw. It was the original version of *Lost Horizon*. If you have ever seen it, you remember that mysterious scene in which Ronald Coleman stumbles out of an airplane that has crash-landed high in the Himalayas. An intimidating party of monks takes him off into the mythical city of Shangri-La. I'll never forget it. The year was 1937, and the plane was the DC-2, the second in a series of three revolutionary new Douglas models that quite altered air transport in the mid-thirties, and which are still in use today.

The story of the Douglas DC-3 began with the death of the great football coach Knute Rockne in the crash of a Fokker trimotor in 1931. His death caused a public outcry over the quality of American air-passenger service. The leading airliners were then the Fokker trimotor, built partly of plywood, and a considerably improved American version, the Ford trimotor, an all-metal airplane with a corrugated metal skin and a fixed landing gear. Both carried about ten people, and both were great machines in their day.

Trans World Airlines (TWA, then called Western Airlines) responded to the outcry. They contracted with the Douglas Company to build an airplane that could take off fully loaded on just one of its two engines as well as beat a Ford trimotor from Santa Monica to Albuquerque. Douglas did just that in 1933 with the experimental DC-1. Then they went into production with the fourteen-passenger DC-2 version and started service with TWA in 1934.

The DC-2 was a fine success, but it was clear to American Airlines that they would have to carry more than fourteen people. They contracted with Douglas, whose chief engineer, Bill Littlewood, wrote the specifications for a third and permanent model of this remarkable new plane. The result was the twenty-one-passenger DC-3, which entered service in 1936. By 1941, eighty percent of the commercial airplanes in the United States were DC-3s, and they were still our most widely used airliner in 1948.

The DC-3 combined variable-pitch propellers and retractable landing gear into a two-engine, low-wing monoplane that was safe, reliable, and easy to maintain. It had an all-metal stressed-skin construction. The aluminum, which replaced the old fabric covering, actually served as a part of the structure. The DC-3 brought us from the airplane design of the twenties to that of the forties in a single step.

None of those advances was unique to the DC-3, but they were brilliantly combined in it. In 1934, a whole array of "flying boats," or sea-

planes, also came into being with most of those features. The flying boats were bigger than the DC-3, and they flew further. Whether they were following Douglas' lead or acting independently, we cannot be sure. But as soon as long-range airliners followed the DC-3, the flying boat's day ended

Nineteen thirty-four has been called the miraculous year of American flight. That year a spate of good ideas emerged all at once. Two years later, the DC-3 formed the almost perfect combination of these ideas. It was a profound landmark within the history of flight. The DC-3 was the airplane that really would take us to Shangri-La.

Another atavistic craving, probably as old as the wish to fly, is the desire to hold on to images—to preserve what we have seen. Art does that, but it does so with massive intervention of the human psyche. We distort the image the moment we lay paint on canvas, pen on paper, or ochre on the cave wall. Might we not find means for saving what we have seen *without* any intervention? We have struggled to retain images as long as we have struggled to fly, and we found means for doing so only a few years after we figured out how to fly in balloons.

We conceived of the camera itself in antiquity. It took a lot longer to invent film to put in the camera. The ancient *camera obscura* took its name from the Latin word *camera*, which meant "a large vaulted room." We get the word *chamber* from *camera*. A *comrade* is literally someone who sits in the same room with us. The camera obscura is a dark room with only one light source—a tiny hole through one wall. That hole projects an accurate image of the outside view, upside-down on the opposite wall. Without film, we cannot really "take" a picture with it, but we can trace the image with a pencil if we want. Aristotle was familiar with the idea, and medieval writers had a lot to say about it.

When I was young, photographic film was comparatively slow, and it came in large sizes. We used the camera obscura idea to make something called a pinhole camera. We would punch a pin-hole in one end of a shoebox, mount film in the opposite end, and seal everything tightly. Then we would point the box at a subject, uncover the pinhole for just a moment, and get a passable photograph.

No one put film in camera obscuras until the nineteenth century, but they were made with lenses as early as the sixteenth century. The term *camera obscura* itself was coined by the astronomer Johannes Kepler,

Major Landmarks

who used one with a fairly complicated lens system to make solar observations in 1600. In the seventeenth century, when the camera obscura was highly refined as an aid to artists, we began to see remarkable improvements in the way painters handled perspective.

Because the longing to capture images is so ancient and elemental, photography did not have to fight for acceptance the way many inventions do. The camera itself had been highly sophisticated for two hundred years, and it was just waiting for someone to invent means for recording the picture automatically. That, in turn, had to wait for eighteenth-century improvements in chemistry.

Thomas Wedgwood, son of the great industrialist Josiah Wedgwood, first recorded images on a coat of silver nitrate. But his pictures were only fleeting; he had no way to fix the image. Two French experimenters, Joseph Niepce and Louis Daguerre, finally created permanent camera pictures in the early 1800s. The engraver Niepce, the first to succeed, spread a bitumen slurry on a pewter plate. Prolonged exposure to light made the bitumen water-soluble. He washed away the soluble bitumen after he exposed the plate in a camera. Then he etched the pewter where it was uncovered. In 1826 he made an eight-hour exposure of the view from his window. That image had taken him ten years of experimentation, but he was the first person in two thousand years to provide film for the camera obscura.

Daguerre was a theater-set designer who made heavy use of the camera obscura in his work. He too was trying to find a way to record pictures. For several years he and Niepce sniffed at each other like wary tomcats. Finally, in 1829, they decided that two heads were better than one, and they formed a partnership. Niepce died three years later, and Daguerre eventually replaced the bitumen with silver iodide and gave us the daguerreotype. The very first daguerreotype process, which dates from the latter 1830s, cut Niepce's eight-hour exposure down to only fifteen minutes, and it produced handsome pictures.

When we look at those first photos we see faces carved by harsher lives than we live today. But even with people in them, the first pictures were carefully composed still lifes—art objects in classical poses. By the 1860s things had changed. Exposures were much faster and attitudes were less romantic. The world was increasingly functional and industrial. Photography turned to documentary reporting—the joining of the two ends of the transcontinental railway, an absolutely radiant young Sarah Bernhardt, a wild-eyed Baudelaire. And another thread appears.

We see rotting corpses of Civil War dead and child laborers in factories. We see the urban poor. By the 1860s photography had become social commentary and a self-conscious historical record. Photography began as an unsolved technical problem, a teaser of the imagination. In just thirty years the ability to record images had turned into an extension of our conscience.[5]

If some landmarks are atavistic, driven by ancient cravings, others represent the awakening to new ways of seeing the world. In Chapter 5 we look at the rise of a whole new means for acquiring knowledge. That change in thinking brought in the new concept of *measuring instruments*. For the first time we sought means for measuring intangible qualities as well as such obvious ones as length and weight. The mechanical clock (Chapter 5) gave the first accurate measurement of abstract time, and it was surely one of the great technological landmarks. The next such intangible quality that yielded to measurement was temperature.

The measurement of temperature ultimately led to refinements of chemistry and to the new science of thermodynamics. To make that measurement required a huge conceptual leap, and the person who made that leap was Daniel Fahrenheit. Fahrenheit was born in the Polish city of Gdansk in 1686. When he was fifteen, his parents died from eating poisonous mushrooms. The city council put the four younger Fahrenheit children in foster homes, but they apprenticed Daniel to a merchant who taught him bookkeeping and took him away to Amsterdam.

There he found out about thermometers. Galileo created an early thermometer and called it a *thermoscope*. Out of that developed the *Florentine thermometer*. Now, sixty years later, the Florentine thermometer showed up as a trade item in Amsterdam, where it caught young Fahrenheit's fancy. So he skipped out on his apprenticeship and borrowed against his inheritance to take up thermometer making. When the city fathers of Gdansk found out, they arranged to have the twenty-year-old Fahrenheit arrested and shipped off to the Dutch East India Company. He had to dodge Dutch police until he became a legal adult at the age of twenty-four.

Left: Galileo's thermoscope.
Right: A Florentine thermometer.

At first Fahrenheit simply went on the run, traveling through Denmark, Germany, Holland, Sweden, and Poland. He studied, learned, and invented. Ulrich Grigull, who tells Fahrenheit's story, points out that Florentine thermometer scales were quite arbitrary. No two were alike. Makers marked the low point on the scale during the coldest day in Florence that year. They marked the high point during the hottest day. Fahrenheit wanted thermometers to be reproducible. He realized the trick was not to use the coldness or hotness of a particular day or place, but to find materials that changed at certain temperatures. Isaac Newton had had the same idea a few years earlier, but he was not a thermometer maker, and his idea stayed in books.

For seven years Fahrenheit worked out an alcohol thermometer scale based on three points. He chose the freezing point of a certain salt-and-water mixture for zero. He used the freezing point of water for 32 degrees. He called body temperature 96 degrees. Why the funny numbers? He originally used a twelve-point scale with 0, 4, and 12 for those three benchmarks. Then he put eight gradations in each large division. That's how he got that 96 number—it was eight times twelve. Body temperature is actually a bit higher than 96, but 96 was close. Later, Fahrenheit made mercury thermometers that let him use the boiling point of water instead of human body temperature for the high mark.[6]

Fahrenheit was still only twenty-eight when he startled the world by making a pair of thermometers that both gave the same reading. No one had ever managed to do that before. The turning points of inventive genius are subtle. Fahrenheit made sense of temperature by seeing temperature scales in abstract terms. He realized, independently of Newton, that scales could be wed to universal material properties. But he also did what Newton failed to do. He built fine thermometers, and those thermometers carried his thinking into the world.

The thermometer was a creature of the seventeenth-century zeitgeist, of course. Modern science was forming a new ability to quantify nature. But if the landmarks of the seventeenth century *served* science, those of the eighteenth and nineteenth *profited from* science. Consider the telegraph, for it represents both an atavistic craving (in this case the wish to be connected to each other in communication) and a leap forward into a new science-based technology that no one previously had imagined. In Chapter 14 we see how the eighteenth-century conception of the telegraph was a product of that century's new fascination with electricity and an idea that occurred long before Morse. The telegraph boldly proclaimed Francis Bacon's scenario of observational science as the servant of invention.

Morse came late in a long series of inventors, each of whom had given us one more piece of a large puzzle. That puzzle was how to use electricity to speak to one another instantly over long distances. I believe it will better help us to understand telegraphy as a major landmark if we look not at Morse but at his contemporary, the almost forgotten William O'Shaughnessy.

In 1855 the British completed a four thousand-mile Indian telegraph system. It connected Calcutta, Agra, Bombay, Peshawar, and Madras. That telegraph was O'Shaughnessy's brainchild, and it secured England's grip on India. O'Shaughnessy had gone to India in 1833 as a twenty-four-year-old assistant surgeon with the British East India Company. There he began experimenting with electricity. He invented an electric motor and a silver chloride battery. Then, in 1839, he set up a thirteen-and-a-half-mile-long demonstration telegraph system near Calcutta.

That was only two years after Samuel F. B. Morse built his famous demonstration system in the United States. But O'Shaughnessy was unaware of Morse's work. His telegraph used a different code, and at first he transmitted the message by imposing a series of tiny electric shocks on the operator's finger. He also came up with another unique invention. He used a two-and-a-half-mile stretch of the Hooghly River, in place of wire, to complete the circuit.

O'Shaughnessy published a pamphlet about the system, but he failed to ignite any interest in telegraphy. Finally, in 1847, Lord Dalhousie took over as governor-general of India. Dalhousie showed real vision in developing public works. He initiated roads, canals, steamship service to England, the Indian railway, and a postal system. And, it was Dalhousie who saw the potential of O'Shaughnessy's telegraph. He authorized O'Shaughnessy to build a twenty-seven-mile line near Calcutta. That was running so successfully by 1851 that Dalhousie authorized him to build a full trans-India telegraph. O'Shaughnessy finished it three years later and it began service in February, 1855.

It was an amazing triumph over technical and bureaucratic problems. By then O'Shaughnessy had learned about the new English and American telegraph systems, but that knowledge was more hindrance than help. It simply meant he had to invent his own equipment to avoid patent disputes. He also had to work with local materials, environments, and methods of construction. He had to invent his own signal transmitter and create his own means for stringing lines. While the system was still under construction, it helped the British in the Crimean War. Three years later, the full system so networked British rule that it was decisive in putting down the Sepoy Uprising. One captured rebel, being led to the gallows, pointed to a telegraph line and bravely cried, "There is the accursed string that strangles us."[7]

Question nineteenth-century British colonialism if you will; there is certainly much to question. But you can only admire O'Shaughnessy. He showed what one person can do by trusting the creative ability that's there for us to claim. He stands as a reminder that one person can make a difference. He also shows us how a true landmark technology wells up everywhere when its time has come. And telegraphy's time had come.

Let us examine one more landmark technology, this one seemingly furthest of all from primitive cravings, but maybe not. This one is the programmable computer, and its story begins with the terribly difficult problem of weaving a pattern into cloth. Different shuttles, carrying the weft strands, have to be threaded through the warp strands in a precise order to give the weave its pattern. In 1805 French engineer Joseph-Marie Jacquard invented means for automating that process. He passed a chain of cards, with holes punched in them, in front of a mechanism that reached through wherever a hole let it and picked up a thread. We have used that principle in textile mills ever since. It is accurate to say that he *programmed* the operation.[8]

Five years later, in 1810, the young Englishman Charles Babbage went to Cambridge to study mathematics and mechanics. In 1816, when he was only twenty-five, he was made a fellow of the Royal Society for his work on calculating machines and methods. In 1834 he made a giant step beyond mere calculation and conceived a machine that could be told how to carry out sequences of calculations. He conceived the idea of programmable computation. He never completed this *analytical engine,* as he called it, but he set down all the essential principles of today's digital computers.

Now, back to Jacquard's loom. The key to operating any digital computer lies in transmitting sequences of on-off commands. Babbage used Jacquard-style punched cards. The presence or absence of a hole communicated a simple on-off command to the machine.

Babbage's idea lay fallow for a long time. Meanwhile, another bright young man, Herman Hollerith, joined the U.S. Census Office—a world of endless copying and tallying. Suppose that someone asked what percentage of our population is made up of Irish immigrants. How do we get an answer from millions of data sheets? One person had tried making ink marks on a continuous paper roll. Then Hollerith thought of punching holes in the paper, like a player piano roll. Holes registered each piece of data mechanically, the way a player piano sounds notes. But that lost the identity of individual records and opened the door to nasty errors.

One day a friend said to Hollerith, "There should be a way to use separate cards with notched edges to keep track of data." Bingo! Hollerith saw it. He developed a system for punching all the data for each person into a single card. If you were a citizen and literate, one hole went in column seven, row nine. He had a full system working in time for the 1890 census.

If you took up the computer before the 1980s, you too worked with Hollerith cards, which happened to be the same size as an 1890 dollar bill.[9] You typed each command on its own card. Hollerith eventually left the Census Office to form his own company. And today that company bears the name International Business Machines, IBM. It is wondrous to see how ideas turn and change and flow—Jacquard to Babbage to Hollerith, and Hollerith's company, at length, building fully evolved Babbage engines for us all to use.

It is all of a piece, of course. For the computer ties into primordial cravings just as surely as flight or communication does. Once the seed of a thinking-machine was properly sown, it grew crazily in all directions, offering every mutation imaginable.

Let us finish this brief exploration of landmark technologies with one more variant on the computer. This machine was consciously conceived as a landmark and today is no more than a forgotten spur in the evolution of computers. It was conceived as a landmark and, in the end, was no landmark at all.

Serious attempts to do complicated machine calculations were under way well before World War II. The most important prewar effort was started in the 1920s by Vannevar Bush, who went on to serve as president of MIT, as a presidential advisor, and as the great technological visionary of fifty years ago. Bush's work on computers culminated in 1942 with the dedication of his huge Rockefeller Differential Analyzer at MIT—a hundred-ton machine with two thousand vacuum tubes and 150 motors. Bush's analyzer was an *analog computer*. *Analog* computers follow physical processes that obey the equations we are trying to solve—in this case, mixed electrical and mechanical processes. *Digital* computers break computations down into sequences of additions and subtractions. They solve equations by doing a whole lot of simple arithmetic.

The Differential Analyzer quickly fell under the pall of World War II secrecy, but only after the head of electrical engineering at MIT had proclaimed it would "mark the beginning of a new era in mechanized calculus" and MIT president Compton had announced it would be "one of the great scientific instruments of modern times." When this wonderful device emerged from secrecy after the war, it had found its way from triumph to obsolescence in only five years. The government had secretly been pouring huge sums of money into developing the ENIAC digital computer to solve artillery fire control problems. The new breed of high-speed digital computers had simply brushed Bush's analyzer aside.

Historian Larry Owens looks at this fall from grace and asks sadly, "How does one tell the story of a machine?" Owens concludes that the real importance of the fall is that it so clearly illustrates a change in the character of engineering after the war. Bush, he says, represented a kind of engineering still in contact with the workshop. His computer was made of complex mechanical and electrical elements. It thought the way prewar engineers thought—in physical, graphical terms. The modern digital computer speaks in a totally different mathematical language to the more abstract and mathematical breed of postwar engineers.[10]

We didn't realize it then, but the failure of Bush's machine served notice that our work as engineers had changed. We were only beginning to see that the Differential Analyzer was almost a kind of android, incorporating those human values of physical intuition and intimacy with process—values that we are trying to rediscover in our work today. The best elements of technology come back in new forms. So if Bush's analyzer is dead, I certainly hope that what it represents is not. The end of the Differential Analyzer was, in fact, the end of the engineering that I was learning in college in the late 1940s. Bush's grand machine served notice that I would spend the rest of my life learning and practicing engineering that I had never even heard of in school.

And so we come back to our question of what a *proper* inventory of technological landmarks would really look like. The answer has to lie in the way such technologies propagate into the future. I would require that any technology on such a list should spread, mutate, and strongly influence the future. That is what all but one of the examples in this chapter hold in common. The majestic Differential Analyzer is the one counterexample, which, I hope, serves to emphasize that point.

12

Systems, Design, and Production

No technology can be reduced to one invention or even to a cluster of inventions. The smallest component of any device, something so small as a screw, represents a long train of invention. Somebody conceived of a lever, someone else thought of a ramp, another person dreamed up a circular staircase. The simple screw thread merges all of those ideas, and it followed all of them. A contrivance made of more than one part is a system woven from those parts. Take a pair of scissors. It consists of just three correlated members—two blades with handles on one end and the bolt that holds them together. Each part represents a skein of invention, and the whole is a device with an efficacy that we would normally not see in the parts alone.

System is a word that takes on new overtones in the modern engineering vocabulary. Yet the modern sense of the word is no different from the dictionary definition, "an assemblage with correlated members." As machines become more complex, however, their systemic characters become increasingly important in the processes of conceiving, designing, and producing them. But the systemic nature of technology does not end with the particular device.

Think for a minute about automobiles. An automobile engine is a large, complex system in itself, but it cannot be designed in isolation from the rest of the car. The engine, radiator, transmission, brakes, air-conditioning, suspension, and much more all act in concert to get you to work or to play.

And the systemic character does not stop there. The automobile interacts with life around it. Questions of service, noise, air pollution, parking, and pedestrian safety all come to rest on the shoulders of automobile makers. That particular assemblage of correlated members reaches even beyond the automobile and its immediate infrastructure. The finished automobile reshapes the society in which it moves. The layout of cities, the design of homes, and even the scaling of the nuclear family have been shaped to this exceedingly complex technology, and that process of change continues still.

As the scope of systems increases, we miss an important feature when we call them mere assemblies of members: the fact that actual members of a given system may not be the same from one day to the next. Houston, Texas, existed in, say, 1850, but then it was a city made up of almost completely different buildings, people, and infrastructure than those that comprise it today. Yet the system we call Houston remains. So a system might better be seen as the mutating relationship among the inconstant parts that make up an assembly.

In any case, a city is just as surely a system as an automobile—a fact that serves to remind us that technology has always had the quality of being woven into the greater systemic fabric of our lives. The scale of our newer technologies has altered the speed and intensity of that interaction. The automobile, telephone, TV, and World Wide Web form systems so large that their engineering absolutely must accommodate large systemic questions.

A few years ago, two of my colleagues in mechanical engineering joined with engineers from other departments in an ad hoc systems seminar. Almost immediately the group expanded to include historians, economists, philosophers, and business administration people. After all, we share the problem of coping with the systems we create. In retrospect, it should be no surprise that the group's attention almost immediately turned to *Gaia*. Gaia is the name given to the biosphere considered as a single living system. A fragile spherical shell of animal, plant, and bacterial life surrounds Earth, extending from the oceans' depths to the outer reaches of the atmosphere. Since all life is interdependent, it forms one living being. We have just begun to understand that to deal with our own occupation of the environment we have to recognize the larger system of which we are only a part.

There, of course, looms the great trap of systems thinking. None of our minds is large enough to grasp the totality of Gaia. Indeed, it is nearly impossible to treat any system when we are not external to it. At best, we learn to move outward from the specific parts we see to some larger whole—in engineering design, in social design, and in caring for our world.

To better understand the complexity of even the simplest systems, try the following simple experiment. First find three metal masses. They might be nuts, bolts, or lead sinkers—whatever is handy. Hang one of them on the end of a thread and swing it. Its motion is that of simple pendulum moving back and forth, or possibly in a circle or a figure eight.

Next, take a longer length of thread and attach all three masses along it. Space them about two feet apart. Then hang this string of masses from the ceiling. Start this system swinging and watch what happens. No matter how they start out, the weights are soon moving in the most unexpected ways. The middle one might momentarily stop dead while the other two gyrate around it. They might all move in the same plane, or they might swing in circles—the patterns of movement keep changing. When we go from one mass to a system of three masses we pass from motion that is complex but mathematically tractable to a motion that is mystifying.

Our technological systems are like that. In October 1987 we saw what happened when a stock market in which much trading was controlled by computers responded to a ripple in the economy. When we had told our computers how to respond to certain market changes, we were completely unprepared for their *aggregate* response. We were stunned to see them flock together and create the greatest one-day stock market crash the world had ever seen. Such a response also occurred during the Three Mile Island reactor failure. So many elements were interconnected that no operator could diagnosis a change quickly enough to take action that would correct the situation instead of making it worse.

Yet complex systems of technologies are at the heart of the machinery of today's society. In 1988 a friend of mine, an engineering designer, came back from Europe and said to me, "John, I had a remarkable experience. I had to call home, so I picked up the phone in my hotel, pushed a few buttons, and found myself talking to my wife in America." I looked at him and said, "So what?"

"Stop and think," he replied. "How many terribly complex systems had to be put together to give me that convenience? There were the space technologies to put up a satellite, electronic technologies on the ground, radio technologies in the sky, hotel management systems, and so on and on!"

So I did stop and think. Today's engineers have reached the point at which they invest far more time in the problems of *combining* elements effectively than they spend inventing them in the first place. Ill-conceived systems threaten us with terrible mischief. Well-combined technologies stand to present us with amazing benefits and conveniences. We have come to the same point in modern engineering that we have with modern medicine. If we try to work on any problem in isolation, when the pieces are all interrelated, we create mischief rather than a solution. We face the same problem that we face with those three metal weights on a string. Their interaction becomes almost incomprehensibly complex.

The problem is compounded as we lose touch with the innards of our devices. The term *black box* has not been in our vocabulary for very long. We first used it to describe any closed array of electronic gear, but it has taken on a new shade of meaning. We now use it any time we have to describe a function that we hide from sight. It has practically turned into a metaphor for our retreat from understanding how things work. Calling the flight recorder of an airliner a black box acknowledges that it is to be opened only under the most dire circumstances.

When I was very young, we stocked radio tubes on the shelf like lightbulbs. When one burned out, we replaced it. Today's radios have transistors. If one fails, we replace the radio itself. Radios are black boxes—I have almost no idea what's in mine. Calculators, car transmissions, and clocks have all become black boxes; even their labels tell us that they can be opened only by factory representatives!

How well do you do in explaining what a carburetor does, what a universal joint is, or how often a spark plug fires as an automobile engine turns over? Few of us know these things today because cars *themselves* have become black boxes. Once upon a time car owners could look right into the transmissions of their Model T Fords. And they had to know how to fix it if they wanted the cars to keep running. The automobile was once a marvelous teacher of applied mechanics. The radio taught a whole generation about electronic circuitry. My grounding in internal combustion, aerodynamics, and electric circuitry came from building model airplanes. Model airplanes took us to technological bedrock in 1943.

Young people today know much that their parents did not know at the same age, but at a price. We handle very sophisticated systems without ever looking inside the systems that surround us. The price we pay is that knowledge itself becomes containerized. A person who knows about computers may not understand anything about cars. John Donne's poetry might remain a black box for a student of nineteenth-century Russian literature. We readily forgive ourselves for ignorance of bodies of knowledge that lie to the right or left of our specialties.

Educating a strong and capable citizenry means teaching students that the black boxes surrounding them are not Pandora's boxes. We must teach students that someone else's subject matter is not a black box, that those boxes can and must be opened. What one fool can do, another fool can also do, and any student is smart enough to open any other student's black box. That in turn brings us back to the matter of systems. Once we realize that we cannot deal with part of a system in isolation, it becomes very clear that encasing knowledge in boxes is one of the most destructive things we do.

So how do we open black boxes when they grow so complex? We need to find the threads of simplicity that run through them. That idea is wonderfully embedded in the old Shaker lyric "'Tis a gift to be simple, 'tis a gift to be free; a gift to come down where you ought to be." Those lines should make up the first chapter in any book on engineering design. It was a lesson I fell into when the army drafted me, two years after I finished college. I was assigned to the Signal Corps Engineering Labs and put to work designing research equipment. There I met a fine designer named Jules Soled—a person who clearly could teach me things. I said to him, "Teach me, and I'll work for you." So teach me he did, many things I hadn't learned in school. His first and last lesson was always this:

Do a first design and then attack it. Your first design will be elegant and complicated, but it will always work better when you get rid of complication. In a really good design you eventually make the very design itself unnecessary. And that is very hard to do because we like complication.

That idea is really quite old. The towering fourteenth-century scholar William of Occam put it this way: "Multiplicity ought not to be posited without necessity." What William meant was that it is foolish to make more assumptions than are really needed to explain anything—

that the simplest explanation is best. We call that idea *Occam's Razor* because it helps slice away the junk in our thinking. Look at the safety razor. For years designers fought with the problem of loading, mounting, and unloading a blade in a holder. If you are old enough, you will remember Schick's "push-pull, click-click" advertisement for its mechanism. Keeping the action workable and the blade solidly in place was a big problem.

Then some bright person applied Occam's razor to the razor-mounting problem and saw that you could simply mold the blade right into the plastic packaging. Now who buys replaceable razor blades? Instead, we set the blades, very solidly and with great precision, right into a cheap throwaway piece of plastic. We have designed blade-holding mechanisms right out of existence.

Gillette Good News double-bladed disposable razor.

We are so tempted to look smart by mastering complication instead of simplicity. Recall that in our Shaker tune, the second line says it is "a gift to come down where you ought to be." Good design exacts a price from our egos, but it really is a gift—it really is freedom—to find the simplicity in things and finally to reduce an engineering design down to where it ought to be.

The road to simplicity can, ironically, be a tortuous and difficult one. Consider an example: Suppose we decide to pass from the obvious form of manufacture, in which we build one item at a time, and instead build parts that can then be put together with parts that someone else has made in another room. Does that represent simplification or complication? In the long run it creates a far simpler form of manufacture, but getting to that point required enormous retooling of our minds as well as of our shops.

The full industrialization of the West took place long after the Industrial Revolution and the rise of steam-powered manufacturing in England in the 1780s. It took until the early twentieth century to finally create assembly line techniques of mass production. The first stage of that process, manufacturing with interchangeable parts, began evolving right on the heels of the Industrial Revolution, but we did not really master it until after the Civil War.

The idea of making machine parts interchangeable is actually quite old. Gutenberg's printing press depended on letters being completely interchangeable. To make printing with moveable metal type into a workable technology, Gutenberg had to couple remarkable ingenuity with a jeweler's skills. When the Industrial Revolution gave us machine tools that could work with high precision, the idea resurfaced. Boulton and Watt managed to make some parts of their early steam engines interchangeable. The idea first entered America when clock makers started making exchangeable gears in the eighteenth century.

Interchangeability had to be far better developed before we could begin mass-producing goods, much less create modern assembly lines. Before the idea could have any utility, we had to be able to create near-identical, machine-made parts without human intervention. Americans like to credit Eli Whitney with first making muskets from such parts in 1803, but that claim falters badly.

To find the first product whose parts could be fully interchanged we go to Paris in 1790. Gunsmith Honoré Blanc had made a thousand muskets and put all their parts in separate bins. He called together a group of academics, politicians, and military men. Then he assembled muskets from parts drawn at random from the bins. By then, Jefferson had already visited Blanc's workshop and written back to

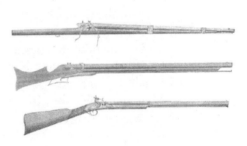

Early handmade muskets, from the 1832 *Edinburgh Encyclopaedia*.

America about the method. Jefferson was president when Eli Whitney duplicated Blanc's demonstration eighteen years later. No one realized it then, but Whitney was faking it. He had carefully handcrafted each part so they would fit together. Whitney sold the government a huge contract for four thousand muskets. He took eight years to deliver them, and then the parts were not interchangeable after all.

So what became of Blanc and his method? The answer is a surprise. For one thing, Blanc had not been first to attempt making muskets this way. Various French makers had worked on the idea since 1720. Furthermore, Blanc went into business, and by the time of Whitney's demonstration he was producing ten thousand muskets a year for Napoleon. Then, in 1806, the French government sacked the whole process. Why? By using unskilled labor, Blanc's method had made manufacturers independent of government control over the old crafts.

The government raised an arcane argument that had a small echo of validity. They said that workers who do not create the entire product cannot produce harmonious products. They simply declared that Blanc's method was not working and they scrapped it.

Meanwhile, America built upon Whitney's promotion of the idea. By 1850 English visitors back from America described what they now called the "American system of manufacture." When they told the French about our use of interchangeable parts, they found the French military had never even heard of it. The French had buried that completely! The story grimly reminds us that technology does not progress in simple, logical ways. Our choices depend on a hundred subjective matters, and they are only thinly influenced by what works best.[1]

Making guns whose parts could truly be exchanged was the large and immediate challenge. The military badly needed to be able to replace a gun part in the field. They never did manage to do that with their service revolvers during the nineteenth century, but muskets and rifles could be made with looser tolerances. By 1860, on the eve of the Civil War, we had achieved interchangeability in rifle parts but were still far from it with the parts of a handgun.

After the Civil War, the idea of interchangeable assembly quickly swept through American manufacturing. The small-arms maker Remington expanded the idea. First he manufactured sewing machines, then typewriters. By the time Henry Ford had carried the assembly line to such a remarkable level in 1913, America was, for a while, established as the world leader in production.[2]

However, it was not Henry Ford who first made an automobile that could be put together with the simplicity that interchangeable parts made possible. Let us change the scene for a moment to England. By 1900 Great Britain boasted that the sun never set on its empire, that Britannia ruled the waves, and that English manufacturing was without peer. Civilization seemed to go downhill in all directions away from London. In 1900 Great Britain took its leadership in powered machinery for granted, but that leadership was built on steam. English engineers disdained internal combustion, and the upper classes saw the development of personal vehicles as a threat to the class system. The best of the early automobiles in England came from continental Europe. England looked at American products with open contempt; no one imagined that America could produce a decent automobile, not even most Americans.

Systems, Design, and Production

In 1903 a young Englishman, Frederick Bennet, worked for an automobile import firm. On a hunch, he imported one of the new Cadillac autos from the United States. He uncrated this six-and-a-half-horsepower, one-cylinder touring car and directly entered it in the Midland Automobile Club Hill Climb in London. Such competitions were teaching a fascinated general public about automobiles. Bennet's little Cadillac chuffed straight up the hill at eight miles per hour and won the competition over cars with three times the horsepower. Then he and his car went on to win other competitions. He convinced people that he had a fine automobile, but he faced another problem when it came to selling the car. If an auto broke down in England, the owner had to send it in to have a new part fitted. Up to then, auto parts were only approximately interchangeable. Factory experts had to file and hammer parts to fit the car. American factories and mechanics were too far away.

So Bennet had to teach the English another lesson. In 1908 he asked officials to select three cars from a set of eight new Cadillacs. Each of the three was publicly dismantled into seven hundred parts. The parts were thoroughly mixed and then reassembled. They fit perfectly into three composite cars. These cars were driven off to the starting line of yet another competition—looking like three circus clowns with their mixed-up colors. Bennet made his point. No one thought cars could be made by mass production with interchangeable parts. These early Cadillacs pioneered the technology, and they opened the British market to American cars.[3]

But more important than that, the cars helped America overcome her inferiority complex. They also opened the American market to American manufacturers. These proud little machines were stalking horses for the fleets of Fords and Oldsmobiles that would transform American life by 1920.

Henry Ford began mass-producing the Model T in 1913. A year later, German submarines began torpedoing Allied shipping. These U-boats, as they were called, were a new threat that had to be fought with new weapons. At first the navy did the best it could with destroyers; then it hired civilian yacht builders to make wooden sub chasers, 110 feet long. They were fast and light, and they looked like yachts, right down to their brass trim. Finally, the navy decided it needed something in between—a two-hundred-foot steel sub chaser called an *Eagle boat*.

In January 1918 the navy hired Henry Ford to build a hundred Eagle boats. Ford was not shy about the challenge. Six months earlier he had told the press with a straight face: "I can build one thousand small submarines...a day."[4] To build Eagle boats, Ford set up three side-by-side assembly lines, each a third of a mile long. He started in May and launched the first boat in July. After that he was supposed to make a boat a day, but things began to go wrong. The first Eagle boat was not ready to be commissioned for three months. A year after Ford began, he had built only seventeen, and their performance record was checkered at best.

Ford had purposely walked around navy expertise. He was not going to be slowed by conventional shipbuilding ideology. After all, he thought, how different is a ship from a Model T? He found there was a big difference. His first boats leaked oil and took on water. Ford's workers had not mastered ship riveting. They got into trouble by using ladders to avoid erecting scaffolds. Ford simply failed to realize how much specific craftsmanship was involved in shipbuilding. The navy cut his contract to sixty boats. Ford finally delivered the last one more than a year after the war ended.

Of the boats that were delivered, all had needed fitting and retrofitting, and even then they functioned poorly. They were awkward at sea, and within five years three had been lost in accidents. When World War II began, only eight of the sixty were still in use, and then only in American coastal waters. A German torpedo sank one of those.

World War II was a new ball game, and Ford turned right around and boasted that he could build a thousand airplanes a day (see Chapter 15). He did build bombers during the war, but nothing like a thousand a day. Planes, like the ships before them, involved specific craftsmanship that had to be built painstakingly into a production line, and Ford's efforts at building planes also failed. Production miracles really had flowed from Ford's self-confidence in the salad days of automobile making. But he eventually lost track of the fact that good technology is a complex fabric woven into people's hearts and imaginations—that a good production line had to embody craftsmanlike knowledge of human beings.

We need to ask ourselves how to recognize industrial innovation when we see it—how to avoid confusing it with simple hubris. How *can* the process of invention and innovation proceed in an industry charged with creating new systems? George Wise offers some clues when he talks about the *product-driven innovation cycle*. Wise suggests that innovation in industry takes place in three distinct ways. The most obvious way is in response to a need. We have repeatedly seen, in earlier chapters, that things are seldom that straightforward—that necessity is no consistent parent of invention. Innovation also occurs when it is driven by some revolutionary discovery. The transistor, for example, triggered all sorts of new technology. But Wise is most interested in yet a third pattern, the product-driven innovation cycle.

It goes like this: An engineer works to improve an existing product, even a modest one. Then he suddenly sees it in a new light, and a radically different concept pops out of it. A butterfly is born of a caterpillar. Wise tells us that sort of thing is far more common than we might think, and he gives several illustrations. For example, in 1908 the General Electric Company hired a young man named Irving Langmuir, an American with a German Ph.D. in applied chemistry. Langmuir's thesis had dealt with the way air carried heat away from small wires. He seemed to be just the person to work on heating elements.

Langmuir went to work on the GE line of stoves, flatirons, and other heaters, using what he knew to improve their heating elements. Now GE also had another product that used heated wires, the lightbulb, so Langmuir looked at it as well. GE's bulbs were all evacuated, meaning that during the manufacturing process all the air was withdrawn from the bulbs so the filament could not burn up. Low-wattage bulbs did well enough at the time, but the tungsten wires in brighter bulbs slowly evaporated. The tungsten vapor was deposited on the inside of the bulbs, eventually turning them black.

Then Langmuir realized that evaporation would be suppressed if the lightbulb was filled with an inert gas that would not burn the filament. The trouble was that gas circulating in a lightbulb removed too much heat and kept the bulb from staying bright. But he knew from his research that a lot less heat would be shed if the filament was wound in a tight coil. By knowing about heat removal, Langmuir gave us the bright argon-filled bulbs we use today. And it all happened because he was put to work on stoves and irons.

Wise traces other connections—how refrigerators led to turbochargers and how steam turbines led to air conditioners. Of course, he is leading us to a common process of invention. A good engineer makes one thing turn into another. He lets innovation be driven by whatever is in front of him. Langmuir had the vision to do just that, and in 1932 he won the Nobel Prize for his work on the surface chemistry of metals.[5]

Now with this impression of systems and industrialization in mind, let us turn to the epoch of the truly grand engineering systems—to the machine seen as a new icon of grandiosity. What the pyramid had been in ancient Egypt and the gothic cathedral had been in the high middle ages, the large engineering system was to become in the nineteenth century. Let us next look at the remarkable age of Heroic Materialism.

13

Heroic Materialism

When Kenneth Clark wrote the concluding section of his magnificent thirteen-part television series, *Civilisation*, he gave it the title "Heroic Materialism." The series had been based on Clark's definition of civilization. Each part displayed an epoch of Western history marked by particular creative energy. Clark finished by showing how, in the early days of the nineteenth century, engineers began building a new world of cast iron—a man-made material world of heroic proportions.

Clark's title was a wry and masterful bit of misdirection, just as the great works he described have also misdirected our attention. If he had used the term *heroic materialism* to describe medieval cathedrals instead of the great engineering works of the nineteenth and twentieth centuries, we might have balked. When we look at a Gothic cathedral we see not cold, material stone, but a flight of the human spirit. We see mind rising over matter.

However, our first reaction is to accept Clark's seeming claim that nineteenth-century iron was merely materialistic. It is a characterization that makes sense to us. But once he has shown the stereotypical view of things, Clark begins steering us in unexpected directions. He shows how all that heroic iron triggered a new spirit of social reform. By the end of the nineteenth century, Victorian iron had even played a role in bringing art back from heroic themes to the humanizing influence of the Impressionist artists.[1]

That should be no surprise since, as we have already stressed, machines relate directly and intimately to essential human needs, and they have always been an equalizing force in society. By the 1930s the Swiss-French architect Le Corbusier was proclaiming machinery and craftsmanship to be the one truth in a world full of lies: "Machines are truly humane, but we do not know machines." He cried, "The world lacks harmonisers to make palpable the humane beauty of modern times."[2] In this chapter we look at the machine as a heroic figure, and what we see is a progression from megalomania to humanization—from the obsessiveness of Napoleon Bonaparte to the world-unifying effort to go into space.

The age of what Clark calls heroic materialism began early in the nineteenth century, and we have shown signs of moving away from that concept of technology only in the last decade or so. The heroic technologies of the last two centuries first mimicked, and then dramatically transcended, our bodies. Today, the impetus to build grand works is just beginning to give way to something else. For today, the computer and modern biology are showing us that great technology is no longer technology that dwarfs us, but that which scales itself to human proportion. Today, great technology mimics our minds as much as our bodies, and it does not so much outrun human thought as rest next to it.

Technology always moves, however haltingly, toward the fulfillment of human need. It has in the past and it will in the future. The technologies of the nineteenth and most of the twentieth centuries enlarged our control over our surroundings. If that process led us to understand that size and control become grotesque at some point, then this too is understanding that, for a time, helped make us more human.

Let us begin, then, on the dark side of nineteenth century technology and visit a little-known side of Napoleon Bonaparte. Two years before he imposed his military dictatorship on France, Napoleon was the twenty-eight-year-old head of the French army. That year, 1797, he was made a member of the Scientific Division of the Institute of France. That's right: Napoleon was honored for contributions to science.

The award was actually much more than a political sop. The young Napoleon was an important supporter of science and engineering. He had done a great deal to strengthen the École Polytechnique, the great French school of engineering and applied science. In 1798, a year after he received his honorary membership, he made his Egyptian campaign into a scientific mission as much as a military one. It was during a military stalemate in Egypt that one of the archaeologists he had brought along discovered the Rosetta Stone.

Heroic Materialism

But Napoleon's support for the applied sciences soon got mixed up with a fixation on architectural monuments. In 1804, the same year he allowed himself to be declared emperor, he wrote, "Men are only as large as the monuments they leave."[3] Historian Frances Steiner tells us that Napoleon dreamed of building monuments from his military spoils—of melting cannon into heroic structures of iron to celebrate battles won. He was still interested in engineering, but that interest had turned to his own glorification.

Bonaparte before the Sphinx, from the 1895 *Century Magazine.*

The snag in Napoleon's dream of immortalization in iron was that ironwork takes a great deal of expertise. The British had mastered it, but France lagged far behind. English iron was expensive, and the quality of French iron was poor. France was still smelting iron with charcoal instead of coke; her engineers had not yet learned the subtleties of building with the metal. Napoleon's new breed of French engineers was eager and surprisingly well prepared to take up the challenge; but French architects were consummate artists in granite. They wanted nothing to do with iron.

During Napoleon's reign as emperor his engineers and architects did execute some major works in iron. They built a number of bridges with varying degrees of success. Once they got the hang of it, the French built a 106-foot arch over the Seine River and named it after the Battle of Austerlitz. The toughest job was using iron to replace a 129-foot dome over a circular grain exchange, the *halle aux blés*. That was finished just two years before Waterloo, at seven times the original cost estimate.

France did not by any means catch up with England during Napoleon's reign. She eventually produced the Eiffel Tower and the structural skeleton for the Statue of Liberty using iron, but that was seventy years after Napoleon. When all is said and done, Napoleon probably did start France on its way to iron construction. But what he gave to technology was something else entirely, something that could never have sprung from a craving for monuments. History has shown that the younger, more idealistic Napoleon really had laid a strong foundation for education in the applied sciences.

The engineers who ultimately did build the nineteenth-century world were driven by new psychic engines altogether. They were melo-

TUNNEL ACROSS THE RIVER THAMES
High Water Line
Low Water Line

Scale of Feet   *Transversal Section of the River at Rotherhithe.*

A cross-section of the far-from-finished Thames Tunnel, from the 1832 *Edinburgh Encyclopaedia*. Note that, at this early date, it was still being conceived as a tunnel for horse-drawn vehicles, not steam locomotives.

dramatic artists in iron, and to understand them, we can do no better than to meet Isambard Kingdom Brunel, the grandest engineer of them all. Isambard's father, Marc Isambard Brunel, was born in France in 1769. He was an engineer and a royalist who fled the French Revolution. He came to America and worked here for seven years. He even became an American citizen. But he finally moved to England to marry a woman he had met in France and known for years. It was his work in England that defined the engineering of the post–Industrial Revolution world. He designed an early suspension bridge, the first floating ship-landing platform, and (boldest of all) a tunnel, the first of its kind, under the Thames River. In the course of digging the twelve-hundred-foot hole they created the first set of technologies for under-river tunneling. They developed one of the first cast-iron shields for protecting the workers as they dug. They had to solve formidable problems of shoring the structure in clay and in sand as those problems arose.

Marc Brunel put his flamboyant twenty-year-old son, Isambard, in charge of construction and they began work in 1825. The task was originally meant to take only three years but it wasn't finished until 1843. During that time, a collapse killed many of the workers, seriously injured the younger Brunel, and halted work for seven years. Yet the completed tunnel has continued to serve London to this very day.

Marc Brunel was solidly creative. But Isambard Brunel brought theater and flair to what his father had begun. He took engineering to the mountaintop and became a nineteenth-century prototype. He built the famous two-mile-long Box Tunnel, major suspension and arch bridges, and a thousand miles of railway. With each project he expanded engineering technique beyond anything previously known or imagined.

His crowning achievements were his steamships. In 1837 he built the paddle-driven *Great Western*, one of the first transatlantic steamboats in regular service. He followed it with an early screw-propeller-driven steamship, the *Great Britain*. Then he bit off a mouthful not even he could chew. In 1853 he began work on the *Great Eastern*, which we have already met in Chapter 9. Designed to take four thousand passengers to Australia and back without refueling, it was seven hundred feet long and weighed twenty thousand tons. The *Great Eastern* was launched in 1858, and Isambard Kingdom Brunel died of stress and overwork the next year. The ship was all it was meant to be, with one catch: it was only a quarter as fuel-efficient as Brunel had expected, seriously blunting its viability as a passenger liner. But it did find its place in history when it proved to be the only ship with the carrying capacity needed to lay the first transatlantic telegraph cable.

The younger Brunel trod the world in seven-league boots of his own making. He made engineering larger than life. He set the mood of the technology of his century. Never before or since have we reached such glorious confidence in our ability to build all the way out to the far fringes of human imagination.

Early in her reign, Queen Victoria and her consort, Albert, saw what was happening all around them, and they hit on the idea of staging a great worldwide exhibition of modern art and design. They created a competition for the design of the central exhibition hall. Sir Joseph Paxton, a botanist and landscape designer, won the competition. Although his building was only a temporary structure, it became one of the great architectural monuments of all time.

Paxton erected his Crystal Palace in 1851. It was an amazing glass-and-iron pavilion, more than a third of mile long, with eight hundred thousand square feet of floor space. The construction had an avant-garde cantilevered iron frame. He made it from interchangeable prefabricated parts and acres of glass panels. It was certainly influenced by the greenhouses he had designed earlier. But he diplomatically claimed that it imitated the organic design of an Amazon lily, *Victoria Regia* (named after Queen Victoria). Isambard Kingdom Brunel had high praise for the Crystal Palace since it was, after all, clearly based on the solid engineering principles he had helped to establish (never mind Amazon lilies).

The exhibition drew over six million visitors and was a huge success until it was taken down in 1854. But it represented a crazy confusion of design styles. Inside, Paxton's array of straight lines was stuffed with the busy rounded artworks of earlier eras: eighteenth-century rococo, turn-of-the-century nature worship—a little bit of this and a little bit of that.

The power of the exhibition lay in the engineering of the building. Victorian art and design lumbered on, ponderous, off the wall, and slightly claustrophobic, while it was Victorian engineering that laid hold of our imaginations. The simple truth was that engineering was the major art form of the mid-nineteenth century. The Crystal Palace itself, not its contents, was the art.

For a generation before the Crystal Palace, the gossamer design of iron bridges had pointed the way to this great iron building. Now Paxton had given all that iron an artistic focus. Conventional painting and sculpture had foundered. For the moment, art was iron. A generation later, as the Eiffel Tower rose over Paris, people such as van Gogh and Rodin would give art a new voice. It took the Impressionists to create art worthy of those iron buildings. And it is fitting that the finest collection of Impressionist art has come to rest in the Musée d'Orsay in Paris—a vast nineteenth-century railway station that is, fittingly, a place of buoyant light and iron.

Even as the Crystal Palace was inadvertently redirecting our attention to the heroic new art of iron, our culture was also being reshaped in copper. After Samuel F. B. Morse showed that long-distance telegraphy was workable, we quickly wove a spider web of lines across America. One of the first was Morse's cable under New York Harbor. Taking telegraphy into the inky ocean depths opened a mare's nest of problems. Still, a cable was run under the English Channel by 1851—fourteen years after Morse's first demonstration.

Three years later, in 1854, an English engineer named Gisborne went to the young American financier Cyrus Field with plans to lay a cable from the United States to Newfoundland. After Field went home to think it over, he came back having decided to go for broke. He set up a company to lay telegraph cable all the way to England. The line to Newfoundland was finished in two years. The waters were fairly shallow, with a silt bottom that protected the cable.

But the twenty-two-hundred-mile stretch under the Atlantic Ocean posed terrible difficulties. The first cables were stranded copper insulated with gutta-percha and tarred hemp. They were wound with three hundred thousand miles of iron wire to protect them and were about half an inch in diameter. No ship was big enough to carry a single twenty-two-hundred-mile length of cable, so it had to be spliced in midocean. The cables broke twice and were lost; but a third try succeeded in 1858.

All the while, scientists and engineers argued about how much voltage it would take to carry a signal over that terrible long distance. The high-voltage people won out, with a two-thousand-volt system. After a month of operation it burned through the insulation at a location just off the coast of Ireland. While it lasted, the cable was met with euphoria. A ninety-eight-word message from Queen Victoria to President Buchanan took seventeen hours to send on the failing cable, but New Yorkers celebrated the linkup with fireworks in the street.

The cable failure was followed by the Civil War, which ended any hope of reviving the project until 1865. Then another failure came to the rescue. The *Great Eastern* may have been less than ideal as a passenger ship, but it was big enough to carry a single twenty-seven-hundred-mile-long strand of one-inch reinforced cable—a single strand that weighed five thousand tons.

The cable broke in 1865, but the *Great Eastern* succeeded in its second try a year later. The public, once bitten and twice shy, was not so excited the second time, even though a stronger cable, operating under low voltage, survived to change the very character of the relationship between America and Europe.[4] It changed much more. It changed the character of life on planet Earth. Some years later, author William Saroyan wrote in his book *The Human Comedy*, "How much does it cost to send a telegram to New Jersey?" The wonderfully apt answer, of course, was, "Not nearly as much as it's worth."

By now France was beginning to catch up in her technologies. Alexandre Gustave Eiffel, born in 1832 in Dijon and trained as an engineer at the École Normale in Paris, was the genius who would lead his country back into the game. He designed bridges and viaducts in his early life and took up architecture later. In 1893 Eiffel was part of a failed French attempt to build a canal across Panama. As an old man, he turned his ever adaptable mind to the new technology of the twentieth century, to flight. He designed one of the early wind tunnels. But all of that pales against his tower. For here Eiffel showed us the humanizing influence of heroic materialism and of iron (in the form of modern steel) as no one else had.

We shall also look at another tower he designed, one even more familiar to Americans than the Eiffel Tower, but first let us deal with the one that carries his name. He built the wild, seemingly mad Eiffel Tower for the 1889 Centennial Exhibition in Paris, and it has marked France to this day. Who can think of Paris without seeing that nine-hundred-foot pylon rising out of its center? Parisians were horrified when they learned what Eiffel was up to. A group of famous writers and artists wrote a manifesto against the tower. It said,

View from within the Eiffel Tower, a quarter of the way up, showing some of the delicate ironwork.

We...protest with all our strength and wrath...against the erection...of the monstrous Eiffel Tower.... This arrogant iron mongery [—this] disgraceful skeleton...even commercial America wouldn't want it.

As Eiffel's monster rose in the middle of the 1889 Paris Exposition, Guy de Maupassant called it "an assemblage of iron ladders," and Léon Bloy said Paris was imperiled by "this positively tragic lamp post springing up from its bowels...like a beacon of disaster and despair."

Finally, there stood that lovely tracery of iron. It is a structure of such subtle delicacy that an exact one-foot scale model would weigh only a third of an ounce! The design is *that* elegant and economical. In 1919 a French writer said, "Great God, what faith its engineer must have had in terrestrial gravitation."

New ideas are alien, no matter how good they are. Parisians' conflicted response to Eiffel's tower should be no surprise. As for Eiffel, from the start he had plainly said what he was doing. He wrote, "The curves of the four piers rising from an enormous base and narrowing toward the top will give a great impression of strength and beauty."[5]

The remark in the Paris manifesto about commercial America turned out to be ironic because, five years earlier, it was Eiffel who designed his other steel tower, the one inside the Statue of Liberty. The Eiffel Tower, for all its grace, did have a hard commercial side, but the Statue of Liberty symbolized both French and American ideals. Liberty is a delicate, graceful copper shell that has been spun around Eiffel's complex but invisible skeleton. She speaks explicitly to the French and American love of freedom. The Eiffel Tower makes no such direct symbolic appeal. Its aesthetic purpose is the structure itself; it is no more than simple beauty in its own right. Eiffel stamped two countries with their identifying marks, and he did so in completely different ways. Of the two, however, the Eiffel Tower signaled the future. Not only did it signal a new concept of beauty, it also carried intimations of a whole new construction, the *skyscraper*.

Now there is a word for you—*skyscraper*! When I was a child, the forty-one-story First National Bank building loomed larger than any-

Heroic Materialism

thing else in St. Paul, Minnesota, and it did indeed seem to scrape the very sky.

For the modern skyscraper to come into its own in the early 1890s, many new technologies had to converge. Before architects could take the bold step of raising buildings beyond about five stories, someone had to develop a workable elevator. Someone else had to recognize that the shell of the building should be hung on a steel skeleton. It was not enough to hang iron facades on wood or brick frames. A whole new technology of building foundations had to be invented. Buildings had to be designed to withstand far greater wind loads than ever before.

The skyscraper was the phoenix that rose from the ashes of the great 1871 Chicago fire. Before the fire, Chicago's architecture was uninhibited and undisciplined. The 18,000 buildings that perished had been made largely of wood and brick, with some iron facing. The commerce that had built Chicago was hardly hurt by the fire, and those commercial interests demanded that a new city be built, a fireproof city, an iron city. There was also pressure to make the best use of real estate in Chicago's crowded downtown area.

Two factions converged on the rubble: the freewheeling builders of old Chicago and a new breed of formal designer trained in analytical mechanics. Many of those new engineers were foreign, and they all reflected the influence of the Napoleonic École Polytechnique. The two factions did not agree, but out of their conflict emerged bold new concepts for making tall steel-framed buildings—concepts that had to be grounded on complex engineering analysis.

These new buildings could not reasonably rise any higher than an elevator could carry its occupants. Hydraulic lifts had been around since the 1830s, but they could not go very high. Elisha Otis invented a safe steam-powered elevator in 1857, but it had to have someone stoking a fire under the boiler. Electric motors would have to be the answer, and the first electric elevators were tried out in Germany in 1880. Practical control systems finally made electric elevators effective in the early 1890s.

By the turn of the century tall buildings were typically fifteen stories high. Thirty years later, the Empire State Building was seven times that height. And the impetus for all this was Mrs. O'Leary's storied cow, who probably did not really kick over a lantern after all. Out of that catastrophe the American city emerged as a completely new invention—a far more extraordinary phoenix than the half-formed city of old Chicago.[6]

If skyscrapers were one dominantly American form of heroic architecture, bridges were another; and for me, the story of American bridges begins in Kentucky, not far from where I used to live. On Sunday afternoons my wife and I would take our kids to see Old High Bridge over the Kentucky River. A plaque credits John Roebling with having started this old bridge in 1853. Actually, the bridge that is there today is a later one that has been entirely reconceived. Still, the glorious spider web of steel emerging out of the quiet hilly isolation around it powerfully evokes Roebling's sense of design and many of his other bridges.

New York City's first skycraper under construction, the 285-foot-tall Flatiron Building, finished in 1902.

The landscape of America is dotted with Roebling's work. Roebling was born in Prussia in 1806. He studied engineering in Berlin, where the philosopher Hegel told him that America was a land of hope for all who were wearied of the historic armory of old Europe. Roebling liked the sound of that and moved here in 1831. First he worked on canal equipment, which led him to invent wire cable to replace the hemp used for tow ropes. Small suspension bridges were gaining in popularity, and Roebling saw that his cables could transform them into something grand. His first contract was to bridge the Monongahela River with an eight-span, fifteen-hundred-foot-long bridge. It was finished in 1846.

Roebling established his name with a suspension bridge over Niagara Falls, finished in 1855. Next he built the Cincinnati Bridge over the Ohio River—a single span more than a thousand feet long. He finished it in 1866, and it is still in use today. While the Cincinnati Bridge was just getting under way, Roebling embarked on his greatest feat—a single-span suspension bridge, sixteen hundred feet long, from Manhattan to Brooklyn. He was hotly opposed by ferryboat operators who stood to lose money and by citizens who thought it could not be done.

He gained financial backing by 1869, but while he was surveying the site his foot was crushed by a loose piling, and he died of tetanus. His son Washington took up the work. It was a terrible task, plagued by accidents and deaths. The paralyzing caisson disease (bends, as we call it today) caught up with Washington Roebling in 1876. It was caused by pressure variations in the huge caissons where piers were being set in the East

188

John Roebling's Delaware Aqueduct. (Photo courtesy of Robert Vogel)

River. The disease rendered him unable to walk or even to talk, so his wife, Emily, took up the on-site direction of the work. When the bridge was finished in 1883, Washington could only look from the window of his house in Brooklyn Heights. He watched President Chester Arthur, his vice president and soon-to-be-successor, Grover Cleveland, and the citizens of New York open the longest suspension bridge in the world.

Today, the Brooklyn Bridge stands as a monument to many things: the vision and determination of the three Roeblings, beauty in design, and of course nineteenth-century ironwork. The bridge has graceful fanlike cross bracing—a pure Roebling touch, and it tells you that you are in New York City as surely as the Statue of Liberty does. It says New York as clearly as the Eiffel Tower says the word Paris.[7]

Another New York landmark is not on the skyline, but it is every bit as heroic a construction as the Brooklyn Bridge. It is the Holland Tunnel, but be warned—its story is a lot like other stories about great engineering works. The plot goes like this: Leaders acknowledge a significant public need. They debate strategy and try to set up a plan for solving the problem. Then a visionary separates himself from the pack. His project is grander and bolder than anyone had expected. He convinces skeptics and puts the project on the road to completion. Finally, he dies creating his masterpiece, and others are left to savor the dream made manifest.

That is how it was not only with John Roebling's Brooklyn Bridge and Isambard Brunel's *Great Eastern* but with the Hoover Dam, the Palomar Observatory telescope, and the Golden Gate Bridge. And it is the story of the Holland Tunnel. In 1906 the island of Manhattan, its population mushrooming, found that it needed more and more avenues of supply. After thirteen years of discussion, a young civil engineer named Clifford Holland was given the job of drilling a highway tunnel under the Hudson River. But such a tunnel faced a brand-new problem in 1919—one that earlier tunnels had not faced. The automobile had come into its own. This tunnel would have to carry forty thousand trucks and cars each day. Vast quantities of carbon monoxide had to be cleared out of it.

Holland began drilling a pair of thirty-foot-diameter holes, almost two miles long, under the Hudson River. The most startling feature of the tunnel was entirely new. Holland provided it with a ventilation system that used four million cubic feet of fresh air each minute to flush out exhaust gases. The project was enormous—a constant string of problems had to be solved, new technology had to be created, and skeptics had to be fended off. It took a terrible toll on Holland. By 1924 he had it under control, but he was suffering from nervous exhaustion. He retreated to a sanitarium, where a heart attack finished him off. He was only forty-one.

President Calvin Coolidge opened the tunnel in 1927. On the first Sunday it carried fifty-two thousand vehicles. Today it carries eighty thousand vehicles per day. Holland's controversial ventilation system, conceived to supply half that number of vehicles, still handles the traffic without trouble.[8] It is an old story, but it does not wear out. For we are always up for one more saga of a heroic technology carried out by a self-sacrificing hero. There are more people to admire, so let us meet yet one more.

This person's name was Elwood Mead, and his medium was dam building. Technology has always seesawed between two poles—carving the world to fit our wants one moment, and yielding to the natural order of things the next. The tension has always been clearest when we try to manage Earth's water. Few ancient technologies touched peoples' lives, or the face of the earth, as strongly as irrigation did. The great Egyptian civilization could be formed only after the arid Nile Valley had been irrigated to supplement the undependable flood waters.

Our early southwestern settlers just as surely had to bring water to the land before they could turn from herdsmen-cowboys into settled farmers. But a great change had taken place by then. First, medieval millwrights coupled water management with power production. Then, during the 1820s and 1830s, French and American engineers transformed the medieval waterwheel into the modern water turbine. By the mid-nineteenth century, water control had been tied to power-generating systems of heroic size.

So the Colorado River caught the minds of engineers as it flowed eighteen hundred miles through the western deserts—two hundred thousand cubic feet of water per second. The Colorado was first explored in 1869. In 1901 engineers diverted a large part of it into a hundred thousand acres of the Imperial Valley in Southern California. The desert bloomed until 1905. Then a flash flood inundated the valley, which was below sea level, and wiped out whole towns. It took two years to regain control of the river. By now it was clear that a huge dam and power plant should be placed across the Colorado River.

Congress debated the project through seven presidential administrations, starting with Teddy Roosevelt's. Finally Herbert Hoover, who had been one of the great engineers of this century, took office. He immediately gave the green light to civil engineer Elwood Mead. Mead began what was at first called Boulder Dam. It became the world's largest concrete structure. Mead saw it through almost to its completion in 1935 and then died just months before it opened.

The dam was soon renamed Hoover Dam, but its reservoir, reaching 140 miles back through Nevada and Arizona, is named Lake Mead. The dam controls water for irrigation and to prevent flooding while it produces over 1434 megawatts of power output.[9] Bigger dams have been built since then, but Hoover Dam opened the way to a whole network of huge hydroelectric dams throughout America. It also represents heroic materialism at its best, for we can see in it the imposition of our will upon the land being coupled with a powerful sense of stewardship toward Earth.

Today, all this delight in size seems to have yielded to a wholly new concept of technology. We talk about microminiaturization and the high technology of the space age. Yet size is still with us. What do you suppose the largest land transport vehicle is? The answer, ironically enough, is directly connected with space flight. It is the *crawler-transporter*, which was created for only one short journey. It is the platform and transport device that NASA uses to carry an assembled rocket and space vehicle on its three- or four-mile trip from the assembly building to one of the launching pads.

This strange vehicle makes more sense as an engineering accomplishment when we realize the magnitude of its task. It is designed to carry a twelve-million-pound rocket and launching derrick and to keep them extremely close to a pure vertical position (within ten minutes of arc) while it negotiates grades of as much as five degrees. The crawler-transporter was selected in preference to either a special barge-and-canal system or a rail system. What finally emerged was something out of science fiction. It is 131 feet long and 114 feet wide. It weighs six million pounds. The structure rides on four double tractor treads—each pair is the size of a Greyhound bus. Inside its huge deck are diesel engines with a total output of almost eight thousand horsepower. They drive generators that supply electric motors for the tracks, the immensely delicate leveling mechanism, the cooling systems, and other internal functions.

A space shuttle approaching the launch site on a crawler transporter. (Photo courtesy of NASA)

The journey to the launching pad requires a highly trained crew of eleven people: a driver, four observers at different locations to advise the driver on steering, and six technicians. The crawler-transporter moves at two miles per hour unloaded and one mile per hour with the rocket in place. Its fuel economy is about 1/150th of a mile per gallon.

And who do you suppose built this high-tech behemoth? Actually, two were built and have served the space program for decades. Yet they were not products of the aerospace industry. They were made by the Marion Power Shovel Company of Ohio—a company with experience in heavy moving technology. The crawler-transporters each cost fourteen million dollars in 1967, when they were put into service.[10]

The space shuttle, however, is about microchips, microecosystems, and microgravity. The contrast between the vehicle and its contents verges on preposterous. A century ago all eyes would have been turned on that huge traveling platform—half as big as a football field. Now we focus on the micromeasurements made in the little space where the humans ride. Who even notices the tractor? Today's technologies create a strange trick of perspective. They cast a mantle of invisibility about the largest machines ever set down on land. Today's vastness lies in dimensions of smallness that open up, layer under layer, when we reach down into molecules and build devices that blur into quantum uncertainty. The big machines hide from us behind their very immensity.

But that crawler-transporter is the true twentieth-century spawn of Brunel, Eiffel, and Roebling. Now our eyes shift from heroic nineteenth-century renderings in iron and take up a technology of speed, space, and light. The rocket reaches for the heavens while we and the crawler-transporter stay gravity-bound. The crawler-transporter calls up all the unbearable contrast of that rocket itself, disappearing into the ether—turning into a tiny golden speck at the very fringe of our vision.

14

Who Got There First?

Years ago, a curator at the Smithsonian Institution said to me, "Scientists and engineers are nutty on the subject of priority." That was before I realized just how far-reaching that nuttiness was or how misguided the very concept of priority is.

As an example, try searching out the inventor of the telephone. Instead of Alexander Graham Bell, you may get the name of a German, Johann Philipp Reis. The common wisdom is that Reis invented a primitive telephone that was only marginally functional, while Bell's phone really worked. Reis was a twenty-six-year-old science teacher when he began work on the telephone in 1860. His essential idea came from a paper by a French investigator named Bourseul. In 1854 Bourseul had explained how to transmit speech electrically. He wrote:

> Speak against one diaphragm and let each vibration "make or break" the electric contact. The electric pulsations thereby produced will set the other diaphragm working, and [it then reproduces] the transmitted sound.[1]

Only one part of Bourseul's idea was shaky. To send sound, the first diaphragm should not make and break contact; instead it should vary the flow of electricity to the second diaphragm continuously. While Reis had used Bourseul's term "make or break," his diaphragm actually drove a thin rod to varying depth in an electric coil. Instead of making and breaking the current, he really did vary it continuously.

Bell faced the same problem when he began work on his telephone a decade later. First, he used a diaphragm-driven needle that entered a water-acid solution to create a continuously variable resistance and a smoothly varying electrical current. Bell got the idea from another American, inventor Elisha Gray. Of course, a liquid pool comes with two problems. One is evaporation; the other is immobility. Bell soon gave it up in favor of a system closer to Reis' electromagnet. Still, it is clear that Gray's variable-resistance pool had pointed the way for Bell.

Next we must ask whether Bell was influenced by Reis' invention. Reis died two years before Bell received his patent. (He was only forty, and he never got around to seeking a patent of his own.) The diaphragms on Reis' phones were too delicate, and his phones were tricky. A German company produced them with inconsistent results. Some worked well. Some transmitted only static. Reis demonstrated his telephones all over Europe, including one that he showed in Scotland at the same time Bell was there visiting his father. We don't know whether or not Bell saw that specific demonstration; however, he could hardly have been unaware of Reis' work.

None of this takes anything away from Bell's brilliance. He produced a robust and viable telephone, and he had the force of personality to sell it to a skeptical public. But to do that, he did what all inventors do. He built on the combined wisdom of others—just as Reis had built on the work of Bourseul before him.

The complex story of the telephone clearly shows how the word *priority* cheats all but one of the many players of credit. In fact, we must thank Bourseul, Reis, Gray, Bell, and many others, because great inventions are always the gift of many people, not just one. Perhaps we write about science and technology the way we do because we so want to tap into the old hero mythologies. A wonderful passage in the apocryphal *Book of Ecclesiasticus* that begins,

> Let us now praise famous men,...
> Such as did bear rule in their kingdoms,
> Men renowned for their power,...
> Such as found out musical tunes,
> And recited verses in writing:
> All these were honoured in their generations.

But then that same passage moves on to a far more realistic, and even poignant, understanding of invention:

And some there be, which have no memorial;
Who are perished as though they had never been.
Their bodies are buried in peace;
But their name lives for evermore....
The people shall tell of their wisdom.[2]

We need not study the history of technology very long before we find ourselves haunted by countless people with no memorial who do, nevertheless, live forever—anonymous inventors of the wheel or the windmill or the plow.

The simple fact is that you and I create our own memorials. If wealth is our objective, then wealth is our memorial. If fame is our objective, fame may very well be all we get. But look around at the memorials of anonymous technologies that have made a better world—leaps of the mind that made the automobile differential, the pencil sharpener, the electric plug, the drop-leaf table, and the lawn sprinkler. I suppose we might find out who invented each of these things, but we are not likely to. Yet the inventions make finer memorials for the quixotic, mentally driven people who gave them to us than wealth or tombstones ever could.

In Chapter 3 we quoted the engineering educator Llewellyn Boelter, who would tell new engineering students, "The products of your minds are the most precious things that you own...you must do the right thing with them." If anything lives forever, it will be those products of our minds, even if they have no memorial. They *are* the most precious things we own. More than that, they are the most glorious things that we have to give away.

Let us then trace some inventions back until they fade into anonymity. Let us tell of the wisdom of some people whose bodies are buried in peace and who have no memorial. And what better place to begin than with the steamboat? Ask anyone in the United States who invented the steamboat, and nine out of ten will give you the name of Robert Fulton. Credit to Fulton for inventing the steamboat is written into the curricula of our elementary schools. It is a "standard fact" of American history.

What Fulton really did was to locate an efficient new Watt engine in a warehouse and, in 1807, install it in a well-designed boat. We were just turning our attention away from England and beginning to think seriously about moving inland. We had a huge network of inland rivers, and people saw that those rivers would provide the means for reaching into the center of our vast continent. Fulton enjoyed immediate commercial success. He had access to a great deal of new technology by 1807, and he put his boat together with an ease that would have been impossible just a few years before. His patent makes no pretense about inventing the steamboat, and it acknowledges thirty years of prior steamboat development.

The earliest successful steamboat that I know of was French, and its story began with two French artillery officers who passed time in camp talking about how they might use steam to power boats. One officer was the Comte d'Auxiron, who left the army in 1770 to work full time on such a boat. By 1772 he had talked the French government into promising that they would give the first successful builder exclusive license to run the boat for fifteen years. So d'Auxiron installed a huge old Newcomen steam engine in a boat. The engine was so heavy that the vessel sank, and after three years of lawsuits, d'Auxiron died of apoplexy.

That might have been the end of d'Auxiron's dream, but while he was just getting started, another young aristocrat, the Marquis de Jouffroy, got involved in a duel. He landed in a military prison on the island of Sainte-Marguerite in the Seine River, the same prison that held the Man in the Iron Mask, about whom Alexandre Dumas wrote his famous novel. During years of enforced contemplation Jouffroy watched boats in the river, and he thought about d'Auxiron.

When he was released in 1775, Jouffroy went to d'Auxiron and his supporters. But he soon decided they had been on the wrong track, and he left Paris for Lyon. There he built his own Newcomen-style engine and, in 1783, made a trial run with a 150-foot boat on the Saône River. For fifteen minutes the boat chuffed past cheering crowds. Then it started breaking up under the pounding of the engine. Jouffroy managed to ease the boat to shore before anyone spotted the failure. He bowed to the cheering crowd and then sent affidavits to Paris testifying to his success. After a long debate, the French Academy of Sciences decided that Lyon never could have succeeded where Paris had failed, and they denied Jouffroy a license.

Finally, the French Revolution drove Jouffroy out of France. He died poor and embittered. Still, he had not failed. After Jouffroy, Fulton could be only aftermath. Yet a great deal more would happen before Fulton, and it is worth our while to tell the sad story of one more early steamboat builder, the American John Fitch.

Fitch was born in 1743 in Connecticut. His mother died when he was four; his father was harsh and rigid. A sense of injustice and failure circled about him from the start. Pulled from school when he was eight and made to work on the hated family farm, he became, in his own words, almost crazy after learning. He fled the farm and took up silversmithing. During the American Revolution he served in the Ohio River basin and spent time as a prisoner of the British and the Indians. Just as Jouffroy had contemplated the Seine River, he contemplated the Ohio, and he came back to Pennsylvania in 1782, afire with a new obsession. He meant to make a steam-powered boat to navigate America's western rivers.

In 1785 and 1786 Fitch, and competing builder James Rumsey, looked for money to build steamboats. The methodical Rumsey gained the support of George Washington and the new government. Fitch found private support, then rapidly built an engine with features of both Watt's and Newcomen's steam engines. He moved from mistake to mistake until he had made our first steamboat, well before Rumsey.

It was an odd machine, propelled by a set of Indian-canoe paddles (see page 106). Yet by the summer of 1790 Fitch had developed it to the point where he could run a passenger line between Philadelphia and Trenton. He logged some three thousand miles at six to eight miles per hour that summer, but in the end it failed commercially. People could not take it seriously. All they saw was a curiosity, a stunt. The impetus to move into America's western rivers was not yet as strong as it would be by Fulton's time, and Fitch (probably because of his mental instability) could not sustain financial backing.

Failure broke Fitch. He retired to Bardstown, Kentucky, and struck a deal with the local innkeeper. For 150 acres of land, the man agreed to put him up and give him a pint of whiskey every day while he drank himself to death. When that failed, Fitch put up another 150 acres to raise the dose to two pints a day. When that too failed, Fitch finally gathered enough opium pills to kill himself. They had called him Crazy Fitch, and they buried him by a footpath in the central square. In 1910 the Daughters of the American Revolution finally put a marker over the spot. It identifies him only as a veteran of the American Revolutionary War.[3]

So this is all the memorial that Fitch managed to

claim. And I am left haunted by the picture of this lonely, troubled man in his beaver-skin hat and black frock coat, the butt of children's jokes, unable to see that his dream had not failed. The steamboat itself honors Fitch far better than he honored himself, for it was he who set the American stage for Fulton. Fitch made it clear that steam-powered boats would be feasible when their time came.

To function creatively we have to function at risk. Watt and Fulton took risks and won big, but not before they had suffered failure. The trick, of course, is to lose one day and come back to win the next. But that is possible only when we draw healthy pleasure and confidence from our creative processes.

The extent of the error of naming a first inventor is illustrated even more dramatically in the case of the telegraph than it is in the case of the steamboat. The noted American painter Samuel F. B. Morse put together a telegraph system in 1837, but his most original contribution in that system might well have been his invention of the code that bears his name.

The seed for the telegraph was sown ninety years earlier, in 1747, when the Englishman William Watson showed that electrostatically generated signals could be sent great distances through a wire, with the circuit completed through the earth. In 1753 an anonymous writer published an article in the *Scots Magazine* showing how it was possible to use an array of twenty-six such wires, one for each letter of the alphabet, to send messages over long distances. Various forms of this multiple-wire system were built in Switzerland in 1774, in France in 1787, and in Spain in 1798.

The notion of sending all the letters on a single wire and using a code to distinguish them was introduced in 1774 (about sixty years before Morse) by a French inventor named LeSage. Still, multiple-wire systems were not completely abandoned for several decades. The whole business got a big boost with the invention of the storage battery. With battery power, people could drive all kinds of information output—such as magnet signals and marks on litmus paper. Between 1800 and Morse's 1837 telegraph, many systems were developed, and many were not bad.

It is worth asking how Morse got the credit he did. His code was the best one up to that time, and his system had the essential features for commercial success, though few of those features were unique. Morse was a man involved in a remarkable range of self-expressive activities—art, invention, politics, photography. He was combative and got into controversies in all these fields. When it came to fighting for telegraph patent priority, he was very effective. In 1854 he won a Supreme Court decision that gave him most of the telegraph royalties. To his credit, he died a wealthy philanthropist. But, as nearly as we can trace it, the *idea* of the telegraph was given to us in 1753 by an anonymous writer. His (or her) reward was not just the fun of dreaming up a wonderful new idea, but having *given* that idea to the world.[4]

Samuel Finley Breese Morse, 1791–1872.

Another inventor whom we should not overlook here is Thomas Edison. If I say *lightbulb*, you will probably think of Edison. Yet the idea of electric lighting was also around long before Edison, and it really got rolling just after 1800—almost eighty years before Edison's contributions. Two different kinds of electric lamps competed with each other all through the nineteenth century. One was the incandescent lamp, where light is created by passing an electric current through a filament, and the other was the arc light.

The brilliant electrochemist Humphrey Davy was probably the first to give us lights of both kinds. The twenty-two-year-old Davy was made a lecturer at the new Royal Institution in London in 1801. He was a dazzling speaker, and his lecture-demonstrations soon became major social events in London for both women and men. In an 1802 lecture he showed that he could cast light by passing an electric current through a platinum strip. In an 1809 lecture he imposed a large voltage across an air gap between two carbon electrodes and created the first arc lamp.

Commercial arc-lighting systems followed three decades later, in England. For a long time, arc lighting was more showy than practical. These systems were just getting good when Edison came along. Meanwhile, in 1820, the French inventor de la Rue made a successful incandescent lamp by placing a platinum coil in an evacuated glass tube. And in 1840 an English inventor named Grove used similar lamps to illuminate a whole theater. But the theater lighting was dim, and its cost ran to several hundred pounds sterling per kilowatt-hour.

Many more incandescent lamps followed. In 1878 an inventor named Joseph Swan made an evacuated carbon-filament lamp three years before Edison did, and he managed to get patent protection before Edison duplicated his feat. When Edison finally installed a complete incandescent lighting system on the steamship *Columbia* in 1880, he provided cheaper, longer-lasting bulbs than anyone else had in a commercially viable lighting system, and his system was complete with an effective electrical supply. To get around Swan, Edison simply took the fellow in as a business partner.[5] Edison's real strength was the tenacious way he developed earlier ideas and drove them all the way into the modern marketplace.

One of the great technologies of our modern world has not been associated with a single titular hero. The automobile is one of those engines of our ingenuity that we really do see fading into a chain of anonymous antecedents. The first steam-powered road vehicles were made in the eighteenth century. Earlier inventors had experimented with cars driven by springs and compressed air. Vehicles powered by windmills were built before them. Leonardo da Vinci sketched self-powered vehicles. And Homer wrote about such machines in remote antiquity.

So let us limit our search to autos driven by internal combustion engines and to autos that were actually built. In that case, the laurel often goes to Carl Benz. Benz believed in the internal combustion engine, and he worked single-mindedly to create an auto driven by one. He succeeded in 1885 and sold his first three-wheeled car in 1887. He went into production with a four-wheeled model in 1890, and Mercedes-Benz automobiles are still very much in evidence today.

Of course, Benz was not really first. A French inventor named Beau de Rochas built both an auto and an engine to drive it in 1862. So, too, did the Austrian Siegfried Marcus in 1864. Marcus's second auto was a pretty solid machine. In 1950 the Austrians pulled it out of the cellar of a Viennese museum and found they could still drive it. It had been bricked up behind a false wall to hide it from the Germans during World War II. Marcus was Jewish, and the Nazis had orders to destroy his car and any literature describing it.

That is as ironic as it is tawdry, because if the German Benz believed in the auto, Marcus did not. In 1898 Marcus was invited as the guest of honor at the Austrian Auto Club. A much older Marcus replied by calling the whole idea of the auto a senseless waste of time and effort.

We are not able to press our search for the earliest internal combustion–powered auto any further back than 1826, when an English engineer named Samuel Brown adapted an old Newcomen steam engine to burn gas and used it to power his auto up Shooter's Hill in London. Yet Benz succeeded where all his predecessors had not. Historian James Flink suggests a reason: Just before Benz made his auto, the modern bicycle had come into being. It had set up the technology of light vehicles, and more important it had sparked public demand for individual transportation. That is probably why Benz succeeded at least sixty years after the first auto was built.[6]

While we are on the matter of internal combustion, the story of Rudolf Diesel's engine plays ironic counterpoint to the question of priority, for Diesel saw himself as the unchallenged sole progenitor of the diesel engine. Historian Linwood Bryant tells us that Diesel felt he was a scientific genius and the James Watt of the late nineteenth century. He was vain, oversensitive, and a little paranoid. He did not win the hearts of other engine makers.

In 1912, twenty years after he conceived his engine, four books were written about its development. Diesel wrote one of them, and people out to minimize his claims wrote the other three. The seeds of the dispute, Bryant argues, were sown in Diesel's view of invention—the usual view that a device is first invented, then developed, and finally improved, all in a linear sequence. Diesel left clear records of what he actually did. There is no doubt that between 1890 and 1893 he invented the engine using his knowledge of thermodynamics. The idea of burning fuel slowly, and at higher pressures, was certainly his.

There is also no doubt that he worked from 1893 to 1897 at the Augsburg Machine Works to develop a working engine. During this time, Diesel faced serious problems. To solve them he had to do a lot more theoretical work and invention. In Diesel's view, he was still inventing the engine. People outside the process viewed this part of the process as development—the dirty work that anyone has to do to make a good idea into workable hardware. After 1897 Diesel figured he was done with his invention, and he turned to promoting it. But the engine was woefully unready for the market. Eleven more years of improvement and innovation were needed. Meanwhile, Diesel worked himself into a nervous breakdown promoting the not-yet-ready engine.

Now the 1912 controversy becomes clearer. Diesel saw his own development as a continuation of the inventive process, and it most surely was that. But the innovation that went on from 1897 to 1908 made the engine commercially feasible. He viewed that period as no more than a time for simple cleanup work by lesser minds. He irritated other engine designers by sneering at their work. He failed to see that what made his engine viable was a lot of truly inventive thinking by many very good engineers.[7]

Diesel was badly troubled by criticisms of his role in creating the

Who Got There First?

201

engine, and in 1913 he vanished from a boat on the way to England. His body was found ten days later. His death brought out all kinds of lurid stories about plots to sell secrets to the British. However, it seems pretty clear that he committed suicide. Like Fitch, Diesel succumbed to his craving for rewards that lay so far from the beauty of his accomplishment as to make us want to weep for his illogic.

The automobile with its various engines thus links us back into a vast history of seeking means for getting from place to place. To make that point, let us look for a moment at *ambulances*. We don't think much about moving the sick or hurt until we have to, and armies have to. That is why armies first created such specialized vehicle.

William the Conqueror's army had an early ambulance. It was a litter carried by one horse in front and another in back. The wounded man rode in a closed box on two poles. He suffered a bone-jarring ride. Various carts and litters show up in military records until Napoleon. Then in 1810 a French army surgeon, Dominique Larrey,

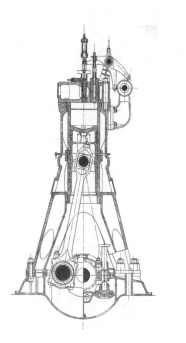

From A.P. Chalkley, *Deisel Engines for Land and Marine Work*. (See note 7)

invented his so-called *flying ambulance*.

Larrey's ambulance was a light two-patient, two-wheeled carrier. It had padding, windows, and some ventilation. It gave a pretty comfortable ride over an open field. The wounded man felt as though he were flying. And that was the military standard through the last century. But then the ambulance as well as its purposes stepped into the twentieth century. To see how, we go to the Smithsonian Institution, where we find an item that looks like an old western chuck wagon with its canvas top and squarish shape. It has a small red cross painted on the side.

This wagon is important. To see why, we go back to Solferino, Italy, in 1859—to a now largely forgotten battle where almost thirty thousand soldiers died in the war for Italian unification. After the battle, a young Swiss named Jean Dunant worked with other bystanders to help thousands of wounded French, Italian, and Austrian soldiers. He wrote a book about that ghastly experience and used it to call for the creation of an international group to give relief in war. The response resulted in the creation of the International Red Cross in 1863. And they chose as its symbol a Swiss flag with its red and white colors reversed.

A few years later the German army started painting the Red Cross symbol on horse-drawn ambulances. But it was in the Spanish-American War that the Red Cross first provided its own ambulances. When Cubans revolted against Spanish domination in 1897, Clara Barton, head of the American Red Cross, asked President McKinley to help her raise public money for Red Cross relief to Cuba. The government finally joined the effort, but only after the conflict had turned into the Spanish-American War in 1898.

Clara Barton, 1821–1912.

The Red Cross raised $36,000. Most of it was spent on eleven mule-drawn ambulances, each carrying four stretchers and a water cask under the driver's seat. Two of the stretchers could be mounted inside as bench seats. The ambulances were made by the Studebaker company—before it began making automobiles. Only two of the eleven ambulances saw action. Those were ones that had gone not to Cuba but to Puerto Rico, where they were very useful. Later, Clara Barton found that the army had not even unloaded the six that were shipped to Cuba.

Two more ambulances were used in New York City, and one saw service with the army at Camp Thomas, Georgia. The army sent that one back to Clara Barton in Washington, D.C., after the war. Eventually a vegetable peddler bought it. The Smithsonian finally located it in 1962. When they restored it, they found that it had originally been painted Prussian blue and chrome yellow.[8]

At first glance this simple ambulance doesn't seem special. Then its meaning becomes clear. You see Jean Dunant's flash of ingenuity after the suffering at Solferino. You see Clara Barton's organizational ingenuity. This humble little wagon represented the first real action by a world relief organization that owed nothing to national interests. In this case, the antecedent of the ambulance is not just a history of prior invention; it is also a vast history of human compassion.

The pace of invention picked up all through the nineteenth century. Toward the end of that singular epoch, the search for priority shifted in quality. Now people sought to get there first. Take the strange story of the *Linotype machine*. From Gutenberg's era until just over a century ago, typesetters had to pick up and set one letter at a time. Starting around 1822 inventors began trying to mechanize this slow process. Ottmar Mergenthaler finally succeeded with his Linotype machine in 1884. After that, Linotype operators set type five times as fast as human typesetters could. But historian Judith Lee tells us about another inventor, James Paige, who patented his own typesetting machine, the Paige compositor, twelve years earlier. Then he joined with the Farnham Company in 1877. The Farnham Company went to its best-known investor, Mark Twain, for support. Twain was intrigued by Paige's machine and began to put money into it. By 1882 Paige had built a functioning compositor.

Paige made two subtle but egregious mistakes in designing his machine. The first was a compulsion to keep improving it. He was not ready to patent the production version until 1887. By then, Linotype machines had been on the market for three years. But Paige was certain he had the better machine. His compositor could set type 60 percent faster than the Linotype. How could he lose?

Mark Twain had long since become a true believer in the Paige compositor, and he had assumed a major financial responsibility for it in exchange for a percentage of the anticipated profits. Then Paige's second mistake surfaced. The Paige compositor was a temperamental racehorse, while the Linotype was a steady workhorse. Paige had designed his machine to function like a human being. He had consciously copied human hand motions. But Mergenthaler had made his Linotype without reference to human function. He understood that machines can move in ways that humans cannot. So his Linotype was simpler, cheaper, easier to maintain, and less liable to break down. Machine tolerances were not as tight.

Furthermore, with eighteen thousand parts, the Paige compositor was far more complicated. It ultimately priced itself out of the market. It took until 1894 for the competitive failure of the Paige compositor to become complete. After that, Paige died penniless in a poorhouse, and Mark Twain went bankrupt. Twain later observed that he had learned two things from the experience: not to invest when you can't afford to, and not to invest when you *can*.[9]

The one surviving Paige compositor is housed in the Mark Twain Memorial in Hartford, Connecticut. It is a beautiful machine, but it is also a reminder that good designs have to do more than just carry out their function. They have to be robust and uncomplicated. Good designs find the solid simplicity that is at the root of anything worthwhile. Thus manufacture and design emerge as parts of the priority question every bit as important as the raw conception of the idea. The soul of the machine is in the details.

That matter comes home to us again toward the close of the nineteenth century, as both Samuel Pierpoint Langley and Orville and Wilbur Wright labored to create powered, controllable flight. Langley worked with government support and enormous public exposure, while the Wright brothers worked quietly using their own resources. Langley attempted flight on October 7, 1903. His huge fifty-four-foot-long flying machine had two forty-eight foot wings—one in the front and one in the back. He launched it from a catapult on the Potomac River, and it fell like a sack of cement into the water. On December 8 he tried again. This time the rear wing caved in before it got off its catapult.

Just nine days later the Wright brothers flew a trim little biplane, with almost no fanfare, at Kitty Hawk, North Carolina. Their advantage was that they had mastered the problem of controlling the movement of their plane, and they had preceded their work with four years of careful experimentation with kites and gliders.

Langley's Aerodrome aboard its houseboat launching platform, from *The Art of Flying*, 1911.

But the government makes an interesting bedfellow. Charles Walcott, a long-time friend of Langley who had been influential in funding his work, was made director of the Smithsonian Institution in 1906—the same year Langley died. He immediately set up a Langley Medal, a Langley Aero Lab, and a Langley Memorial. Then, in 1914, he funded Glenn Curtiss, who had been involved in a bitter patent dispute with the Wrights, to reconstruct the Langley machine and show that it really could fly.

Curtiss went to work, strengthening the structure, adding controls, reshaping it aerodynamically, relocating the center of gravity—in short, making it airworthy. In 1914 he flew it for 150 feet, and then he went back to replace the old motor as well. On the basis of Curtiss' reconstruction, the Smithsonian honored Langley for having built the first successful flying machine.

In 1925, Orville Wright at last roused American opinion to his cause by placing the original airplane, this American treasure, in the Science Museum of London. In 1942 the secretary of the Smithsonian, Charles Abbot, finally authorized publication of an article that clearly showed the Langley reconstruction was rigged. Orville responded by telling the British that his airplane should be returned to the Smithsonian Institution after the war. He died in January 1948, and eleven months later the first airplane returned to America—to the Smithsonian, where it now hangs over a label giving the Wright Brothers their due.[10]

Today, of course, Langley's name graces a major NASA center, a military air base, and the CIA headquarters. Justice does indeed seem to have prevailed for the Wright Brothers. But the simple fact that Langley played a role in getting us into the air also seems at last to have become a clear fact of history.

Once we had found our way into the air, another kind of priority dispute came to the fore: the competition for records. Surely the most notable of those record quests was to be first to fly the Atlantic Ocean. And so by the time Lindbergh made his historic nonstop flight from New York to Paris in 1927, dozens of people had flown the Atlantic.[11]

The first such flight was made in May 1919 from New York to Plymouth, England, in the navy's six-man, four-engine NC-4 flying boat. But it stopped in the Azores and Lisbon on the way. That same month Raymond Orteig of New York City offered a $25,000 prize for the first nonstop airplane flight from New York to Paris. Just one month later, Alcock and Brown flew a two-engine airplane nonstop from St. John's, Newfoundland, to Clifden, Ireland.

In July 1919 a British dirigible flew from England to New Jersey and back, and in 1922 two Portuguese aviators, Cabral and Coutinho, flew a single-engine British seaplane from Lisbon to Rio de Janeiro. That's a longer flight than Lindbergh's, but there is a catch. The flight did not just involve a stop—they actually changed airplanes on a small Atlantic island. More New York–to–England flights followed in 1924. And in 1924 a zeppelin flew from Friedrichshaven to Lakehurst, New Jersey. Finally, in 1927, seven transatlantic heavier-than-air flights were made, of which Lindbergh's was the third.

So what was special about Lindbergh's accomplishment? Well, it was the longest nonstop heavier-than-air transatlantic flight and it was the first solo crossing. That's how he picked up the nickname of "The Lone

Eagle." And his flight finally fulfilled the specific conditions of the Orteig prize, which had by then been without a claimant for more than eight years.

Prevailing headwinds made it a lot harder to fly from Europe to America. The first solo heavier-than-air flight from east to west was not made until 1932. The pilot's name was James Mollison, and he flew only from Ireland to New Brunswick. It was 1936 before Beryl Markham finally flew east-to-west from England to the Canadian mainland. (She wrote about it in her wonderful book, *West with the Night*.)[12]

The NC-4 airplane, which made the first flight across the Atlantic, and its pilot, A.C. Read, from the 1923 edition of *The World Book of Knowledge.*

Commercial transatlantic flights had to wait until the late 1930s—about twenty years after the first transatlantic crossing and thirty-five years after the Wright brothers. Still, it was Lindbergh's flight that riveted public awareness, and it is worth looking at the airplane he flew. Lindbergh was a determined pilot who finally found a like-minded designer in the owner of the tiny Ryan Airplane Company. Ryan specially built the *Spirit of St. Louis* in just two months for Lindbergh. Lindbergh was also called "Lucky Lindy," and it seems that a good part of his luck was having found the right engineer at the right time.

Perhaps the best view of how treacherous questions of priority become was given to us by Thomas Kuhn in his 1962 bombshell account of scientific change, *The Structure of Scientific Revolutions*. Kuhn showed that truly revolutionary changes in scientific understanding were the work not of single geniuses, but of new scientific communities that gathered themselves around wholly new scientific outlooks. One of Kuhn's most dramatic illustrations is his account of the "discovery" of oxygen.

It is common knowledge today that the oxygen in air reacts with other materials when they burn. But eighteenth-century chemists thought burning materials were simply releasing an invisible fluid called phlogiston, which caused heating. No one supposed burning had anything to do with the air itself. They had no idea that the active agent was oxygen, which makes up one-fifth of air. Oxygen was finally pinned down as a separate element by three people in the 1770s: an English cleric named Priestley, the French chemist Lavoisier, and a Swedish pharmacist named Scheele.

When Priestley isolated oxygen in 1774 he first thought he had laughing gas. A year later he decided that he had actually removed the phlogiston from air. At the same time, Lavoisier (who knew about Priestley's work) also isolated oxygen. He took it to be very pure air.

Two years later, Lavoisier realized that he had actually separated a component of air; but he thought it came into existence only when the air was heated. Meanwhile, the Swede, Scheele, had been working quietly. He published a book titled *Air and Fire* just after Lavoisier uttered his final word on the matter. In it, Scheele identified oxygen as a separate part of air, based on work he had done before either Priestley or Lavoisier.

Kuhn uses this muddle to completely rewrite our concepts of priority. Squabbles over credit, he explains, cloud the real nature of discoveries. Should we credit Priestley, who isolated oxygen and then went to his death thinking it was something else? Should we credit Lavoisier, who saw it as part of air but failed to understand its nature? And what about Scheele, who published his work too late in the game?

Oxygen could not really be understood in terms that we find acceptable until scientists changed their whole view of matter. Priestley started a scientific revolution that would not be finished until John Dalton built oxygen into the atomic theory of matter thirty years later. The idea that burning meant new combinations of atoms was too great a leap for any one person to make. The pieces of the puzzle added up until suddenly an unexpected new picture became clear. Oxygen was not *discovered*. Oxygen as we understand it today could not have been perceived, much less discovered, until a whole new science was forged to accommodate it. Priestley, Lavoisier, Scheele, and Dalton each added new insights that finally forced a major scientific revolution.[13]

The situation is not much different in engineering and technology. Inventions burst forth when the world is ready to shift under our feet. We may designate this or that contributor as the titular inventor, but that is only a matter of convenience. Most of the people who give us new technologies have no memorial. Rather, we select arbitrary heroes and let them tell of the collective wisdom. We celebrate Edison, Bell, Morse, and Fulton not because they *transcend* you and me but because they *encapsulate* what you and I are capable of doing if we choose to. Those arbitrary heroes reflect what countless others (like us) have already done, over and over, to put us where we are.

15

Ever-Present
Dangers

Amurderously recurrent theme surfaces as we read the record of
technology. It can be decocted into the tidy epigram: "The
fastest route to success is through failure. The greatest enemy
of success is success." When my civil engineering colleague Jack Matson recognized the validity of that idea, he began vigorously to promote
the concept of *intelligent fast failure*. He said that we can speed our own
creativity if we begin by running through as many wrong or foolish
ways of accomplishing our end as we can think of. That process both
emboldens us and instructs us in the full range of possibility. Con-
versely, success that fails to keep the boundaries of error within sight
eventually takes itself for granted and leaves us open to failure on a
grand scale.[1]

We skirted this issue toward the end of Chapter 9; now let us look at
it more closely. A story of three bridges helps to expose the complex
way in which success and failure work together. Henry Petroski takes us
back to the forty-six-mile rail trip from Edinburgh to Dundee, which
took half a day in 1870. Passengers had to ride the ferry over two wide
fjords, arms of the North Sea slicing into Scotland. They are the Firth
of Tay and the Firth of Forth. Then an English engineer, Thomas
Bouch, sold backers on the idea of building bridges over those inlets.

The first was an immense two-mile bridge over the Firth of Tay.
When its eighty-five spans were finished in 1877, they made up the
longest bridge in the world, and Queen Victoria knighted Bouch. Disaster followed almost immediately. The Tay Bridge collapsed in 1879,
killing seventy-five people. Cost-cutting had yielded a bridge that

couldn't stand up to the wind forces. Bouch died in humiliation four months later. By 1881 the Tay Bridge had been rebuilt with heavy, unbeautiful trusses, and attention turned to the second bridge, the one over the Firth of Forth.

The Firth of Forth bridge was to cross where the center of the firth was a mile wide, with only one shallow spot for a central pier. That meant the center portion of the bridge had to be built in two huge half-mile spans. How to proceed? An engineer named Benjamin Baker rallied public support for a new kind of cantilever construction, with triangular cross bracing. The cantilevers reached out from opposite sides to hold the central spans. That was utterly radical engineering in 1881. When Baker's bridge was finished, it was a stunning array of massive tubular steel members.

1920s stereopticon photo of the bridge over the Firth of Forth.

The third bridge was built in Canada. While Bouch and Baker struggled to link Edinburgh and Dundee, a young American bridge builder, Theodore Cooper, watched and learned. He was forty-eight years old and a seasoned designer just opening a consulting office in New York when the first Tay Bridge fell. Cooper was known for hands-on care in his work. He knew not to be careless the way Bouch had been. He knew to stay on top of the work.

In 1887, as Baker's Firth of Forth bridge was going up, the Canadian government set out to bridge the St. Lawrence River at Quebec. The St. Lawrence was yet another wide inlet, one that split eastern Canada in two. And, like Baker, Cooper designed a cantilever bridge. But years of success had buoyed his belief in saving money on material. His Quebec bridge was light and buoyant, and it was to be the longest cantilever structure ever made. By 1907 its delicate girders reached, unsupported, nine hundred feet over the water.

Cooper was now sixty-eight years old and America's leading bridge designer. He had grown content to sit in New York directing the work by telegram. When a message warned that beams were buckling, he failed to react in time. The arm collapsed and seventy-three workers fell to their death.[2] So history repeated itself—success following failure and failure following success in a cycle we forget at our peril. Today, all three inlets have their bridges. But, of the three, it was the radical Firth of Forth bridge, built in disaster's wake, that *didn't* suffer a collapse—the only one of the three that was built in the right phase of that murderous cycle.

Engineers have always ridden that cycle. Four and a half millennia ago, the Babylonians began evolving a body of law that included the matter of technological failures. (A part of that body of law was carved on a large stele during the reign of Hammurabi in the eighteenth century B.C. The stele was rediscovered in 1901.) The so-called Code of Hammurabi says little about punishing murder, but woe betide anyone guilty of *failure*! The surgeon who bungles an operation loses a hand. The mason whose building collapses is punished on a sliding scale: If the owner is killed, the mason himself is executed. If the owner's son dies, then so must the mason's son, and so forth. The code is not written for the squeamish. It specifies punishment by amputation, impalement, drowning, immolation, and enslavement—all with blood-chilling abandon.[3]

Getting things right is a far bigger worry in today's dense technology than it was thousands of years ago. Yet while we do not threaten to amputate surgeons' hands or kill engineers' children, our resulting technologies are still surprisingly safe. Only one person in ten million dies each year from the structural failure of a building. And tens of millions of Americans safely make commercial flights between the rare fatalities that do occur.

Our equation for preserving safety is different from that in the Code of Hammurabi. A world that depends on technology has to allow some risk. Trying to function in perfect safety would make technology hopelessly static and would rob it of all vitality. I feel gravely cynical when anyone speaks to me of zero defects or zero tolerance. What we do is to set the levels of acceptable risk and then punish the engineer or doctor who is reckless.

Of course, we are seldom evenhanded about it. We accept familiar kinds of accidents but impose extremely high standards on unfamiliar technologies. Down through the twentieth century we were remarkably casual about deaths from tobacco, coal mining, and drunk driving, while we went to the wall over less familiar technologies (such as nuclear power) that killed almost no one.

We have to look clearly at risk. Threatening to chop off an inventor's hands has never been a useful strategy for improving technology. Still, we can hardly be casual about the terrible dangers each new technology poses. I have no Hammurabic formula for striking the balance, but in a complex world balances must be struck. Of course, the underlying problem is the widening gap between technical literacy and the things we have to know in a technological world. We are heaped with technical information. Some of it is accurate; some of it is terribly wrong; some of it is only innocently off base.

A mechanic tells me that a part in my car failed because it crystallized. That's because fatigue failures run along crystal boundaries in metals. Failure reveals a crystal structure that was there all along. I suppose there is no real harm in talking about crystallization. But we are surely better off knowing that metal does not mysteriously change its elementary structure with age.

I was more worried by an ophthalmologist who said I should have laser surgery. A laser, he explained, carries both light and heat. Actually, lasers are used in retinal surgery because they carry *no heat whatsoever*. The energy of a laser beam is pure coherent light. It passes right through the optic fluid. It is transformed to heat only when it hits the retina. But optic fluid *absorbs* heat radiation. If there were any heat in a laser, it would be absorbed in the eyeball instead of the retina. The consequences of that are too horrible to consider.

We run into the similar misconceptions from some aluminum-siding and storm-window salesmen. It is fascinating to imagine a physics book based on some of their theories. But who am I to complain? As a joke, I start one chapter in my heat transfer textbook with a tall tale. It goes like this:

> When I was a lad, winter was really cold. It would get so cold that if you went outside with a cup of hot coffee, it would freeze. I mean it would freeze *fast*. That coffee would freeze so fast that it was still hot after it froze. Now that's *cold!*

When someone looks me straight in the eye and says, "Gosh, I didn't know it could get that cold," then I have to repent my own sins.[4]

We live in a technology-dense world. The engines of our ingenuity are everywhere, and we are terrifyingly naked without knowing elementary things about how they work. A good citizen necessarily knows a little about the flow of electricity or of fluids—about chemical reactions or the insides of automobiles. A world unknown is a world we cannot cope with. It is not a nice place to live. We have all heard people say, with authority, that hot water freezes faster than cold water, or that storm windows let heat flow only one way. We are in trouble if we have no means for at least testing statements such as these. None of us can know everything we would like to know, but we must be well enough educated to be proper skeptics. We need enough knowledge to question the things so many people try to tell us.

Let us look at one surprising and terribly important counterintuitive attribute of the physical world, which is already so permeated with issues of safety and danger, success and failure. It is the slippery concept of *stability*. To understand stability you might do a small experiment. First, try to balance a long pencil, on end, on your fingertip. If you can manage that (and I gravely doubt you can), then try to lift your finger, keeping the pencil balanced as you do. It takes precisely that kind of balancing act to launch a rocket. The main nozzle pushes the rocket upward. Smaller nozzles direct jets off to the sides. When the rocket starts to tip over, one of these jets fires to realign it, just as you try to do by moving your finger to realign the pencil.

The rising rocket is a completely unstable system. Left to its own devices, it will always tip over. When I was a kid building model airplanes, I knew that some designs were stable and others were not. A World War I biplane looked nice on the shelf. But without a pilot at the controls to make corrections it would stall, fall into a tailspin, or find some other way to crash. On the other hand, models of gliders or of Piper Cubs flew wonderfully well when I set them loose in the air.

Throughout the history of flight, airplane designers have entertained very different views of stability. Most nineteenth-century builders were trying to invent stable flying machines, and none succeeded. But the Wright brothers' airplane was unstable. They knew the trick was not to make the machine stable, but rather to make it *controllable*.

It helped that the Wright brothers were bicycle makers who understood that a moving bike is unstable—that when we quit steering we fall over. At the same time, a bicycle maneuvers perfectly well with constant minimal control. Instability was the heart of a World War I biplane. The very instability of those old flying box kites made them highly maneuverable. No stable airplane could do a chandelle or an Immelmann turn. Not until flying went commercial in the 1930s did designers take much interest in stable airplanes. Commercial flight could be very dangerous if it had to have constant, close control. Commercial airplane designs remained stable until the modern jet plane. Modern jets are highly unstable, but they are equipped with sophisticated automatic controls that give the pilot an *illusion* of stability. Today we've gone back to unstable designs that need a great deal of control. But humans need do only a small part of the work of controlling them.

The issues of stability and of Hammurabi's code converged in Russia in the 1930s. In Chapter 10 I stressed that kings and emperors are outside forces that cannot interfere with the creative process and that their wars do a very poor job of driving technology. Joseph Stalin made that crystal clear after he completed his takeover of the Soviet Union in 1929 and began a ruthless program of collectivization and purges. In 1933 he started a campaign to build up Russian morale and draw attention away from his ongoing slaughter of so-called enemies of the people. He flung Russian airplane designers and pilots into competition for international flight records. That way, the Russian newspapers were able to boast of technical success after success, while Soviet citizens were being trucked off to the labor camps.

Russia first entered the record books in January 1934, when three Russian balloonists bested an American altitude record (and died doing so). The year before, Stalin had seized on the work of the famous Russian designer Tupolev. Tupolev was already developing a long-distance airplane, and by 1938 Russia had claimed some sixty-eight records for distance, altitude, and various other firsts. One of the more spectacular ones was a sixty-three-hundred-mile polar flight from Moscow to San Jacinto, California, in 1937. Before each flight, Stalin met the pilots, discussed their plans, and publicly worried about their safety. He met returning airplanes while flashbulbs popped. All the while, the death toll among Russian pilots rose. Then two things happened.

First, Russia's lead started to slip. In 1939, for example, a plane left Moscow seeking to make a record-breaking flight to New York. It was to arrive in time for the opening of the New York World's Fair, but it crashed in New Brunswick, Canada. The pilots eventually arrived in New York in an American rescue plane.

Russia's greater failure came in the Spanish Civil War, which became the ghastly proving ground for both fascist and Bolshevik ordnance before World War II. By 1937 it had become clear that Russian airplanes, designed to win distance and altitude records, were no match for German combat planes. The lumbering long-distance Russian airplanes were stable in flight. German airplanes were fast and maneuverable as only very unstable airplanes could be. Stalin reacted by imposing his own Hammurabic code—he jailed Tupolev and nearly five hundred of his aeronautical engineers. Russian aviation had done well in the short term, but it never did recover from the long-term damage that Stalin had done by interfering with the free flow of ideas.[5]

On this side of the Atlantic, a similar mischief was afoot. In the gathering days of World War II our air force badly needed lots of airplanes, so the military went to the automobile manufacturers for help. In 1941 General Motors started making B-29 bomber parts, and Ford set up a plant to make B-24 Liberator bombers at Willow Run. Then General Motors hired Don Berlin, the man who had designed the P-40. The P-40 was the fighter plane that the Flying Tigers had been using in China.

Up to then, airplanes had been virtually handmade. Historian I. B. Holley says that the biggest airplane company could turn out three planes a *day*, while automobile makers made three cars a *minute*. Henry Ford, for one, had no appreciation for how hard it was to adapt a good airplane design to his mass-production methods. He rashly claimed that, freed of government red tape, he could make more than a thousand airplanes a day. That, of course, echoes the Eagle boat story in Chapter 12. Unfortunately, it echoes the Eagle boat story in its outcome as well.

Don Berlin went to the air force with a plan that would link his skills as an airplane designer with the General Motors style of mass production. He showed them a new fighter plane design, the XP-75. It was supposed to outperform even experimental fighters. But he was going to avoid the problems these planes were giving us. His plan was simple. He would use the best parts of other airplanes: the wing of the P-40, the tail of the A-24, the landing gear of the P-47, and so on.

Like the creations of Pygmalion or Frankenstein, the XP-75 was to be an assembly of perfect parts. Unfortunately, the result resembled Frankenstein's creature in more ways than one. It was an oversized monster that could not begin to compete with planes designed from the ground up. The air force was so sure this beast would succeed that it went ahead and tooled up to mass-produce it. When the whole thing was finally scrapped in 1944, it had cost over nine million dollars. That was still a great deal of money back then.[6]

A designer must, after all, seek out the harmony of the many parts that make up a design. A designer must see the design whole. Sophia Loren once pointed out that all her parts were wrong. Her nose was the wrong shape, her mouth was too wide, and so on. Yet who could fault her beauty? The components of a good design have to make sense in the context of the whole. The XP-75 was no more than a collection of fine parts. It was not a whole airplane.

The matters of stability, failure, and government intervention rise again in the story of the early American satellite *Explorer I*. The first two satellites were the Russian *Sputnik I* and *Sputnik II*. They went up in October and November of 1957. Our *Explorer I*, launched in January 1958, was third. The two Sputniks were almost spherical, but *Explorer I* was long and narrow, like a pencil.

Explorer I was meant to rotate around its own center-line like a pencil spinning about its lead. It was definitely *not* supposed to rotate end over end like an airplane propeller or a windmill blade. Technically speaking, a pencil spinning about its lead is in its minimum-moment-of-inertia mode. The windmilling pencil is in its maximum-moment-of-inertia mode, and it spins more slowly.

A radio astronomer named Ronald Bracewell at Stanford University tracked the first Sputnik and determined that it was spinning in its maximum-moment-of-inertia mode. As its antennae flexed, it dissipated a small amount of rotational energy. That destabilized any rotation that was *not* in the maximum-moment-of-inertia mode and shifted its rotation, but it made no difference in the case of the almost-spherical Sputnik. Bracewell

Explorer I. (Image courtesy of NASA)

knew about the behavior because galaxies behave the same way. What he knew, and Explorer's engineers did not, was that *Explorer I*'s rotation would be unstable. It would soon flip over and start windmilling through space. That's what a spinning coin does. It starts out spinning about one of its diameters. Then, as it spends its energy, the rotation flattens out. It tries to go horizontal and spin like a turntable.

So Bracewell called engineers at the Jet Propulsion Laboratory to warn them. But the people in charge of security wouldn't let him talk with the engineers. He had to get the word out by publishing a paper in the open literature. The paper came out seven months after *Explorer I* was launched. Once up, *Explorer I* made just one Earth orbit. Then it flipped over and, from then on, windmilled across the heavens.[7]

Bracewell's story might have ended much differently in today's Internet-served world. So too might the story of another engineer named Landon who worked at RCA. Before *Explorer I* was launched, Landon described in unpublished laboratory notes the kind of instability suffered by *Explorer I*. Of course, information that doesn't flow freely is really not information at all. There can, I suppose, be some acceptable reasons for secrecy in technology, but make no mistake: Secrecy is an enemy of progress. Creativity, freedom, and openness are natural bedfellows.

Before the Berlin Wall came down, a Russian engineer pointed out why the United States stayed ahead of Russia in computer development. Once we had established a lead, he said, Russia tried to keep up by copying what we had already done. They had plenty of ways to break through our security. But, forced by their system to play a copycat game instead of trusting their own inventive genius, they were trapped in a technology that was doomed to stay one step behind us.

Perhaps hubris is what gives success such destructive power in our lives—the sense that we can stop listening. Hubris of one kind brought down the Quebec bridge, as we've seen. Other forms of hubris denied air power to the Russians and gave us the XP-75. The failure of *Explorer I* could have been avoided if government security people had taken themselves less seriously and been willing to listen.

Another terrible failure, the collapse of the skywalks in the Kansas City Hyatt Regency Hotel in the summer of 1981, could also have been avoided if construction people had been attentive to one another. Disaster struck when a weekend crowd was dancing to big-band music in the hotel atrium. Some people were on the ground floor; others strolled on three crowded skywalks above the atrium. Two of the skywalks suddenly gave way, and 114 people were killed and two hundred more were injured.

Twentieth-century thinking being what it is, the first order of business was to fix blame—to find out who was entitled to sue whom for how much. The National Bureau of Standards issued a three-hundred-page report one year later. It leads the way down a tortuous path through a sequence of three minor complacencies on the part of otherwise honorable people. The walkways were hung from the ceiling on steel rods. On one side of the atrium, one walkway had been suspended below another.

In the original design, these two walkways were to have been mounted on the same rods. Even at this stage of the design two errors had already been made, neither of which would have been fatal. The first error was that the rod design failed to meet the building code. While that made the design illegal, a stress analysis showed that the rods were still safe. The second problem seemed to be only a small detail. The designers had not been clear about how the rods were supposed to grip the upper walkway where they passed through it.

The contractor then unwittingly made the third error of judgment. He solved the problem of gripping the walkway by simply ending one rod and starting another next to it in the skywalk's crossbeam instead of using one continuous rod. The result was subtle but devastating. This change doubled the stresses at one point in the beam. Although the effect is subtle, the calculation is still within the grasp of second-year engineering students. (In 1988, as a homework exercise, an applied-mechanics professor at the University of Houston asked such students to calculate the stress that resulted from splitting the rod.)

So, under extreme loading, the upper walkway failed and both walkways fell away, depositing steel, concrete, and people upon the dancers below.[8] It was a terrible moment. But it never would have occurred if the chain of blunders had not fit together so perfectly. I am chilled by the weight of responsibility that rides on the communication of details among the many people who execute a design. At the same time, I am encouraged to see that so much safety is inherent in our system of design—that so many dovetailing errors had to be made before this dreadful accident could ever happen.

And so, again and again, success breeds complacency breeds failure breeds caution breeds care for the details breeds success breeds hubris breeds insensitivity to our fallibility breeds oversight breeds failure breeds a broken spirit breeds renewed care breeds....

It takes remarkably strong character for any technologist to stay out of that cycle. But some do. The good news is that some people are able to stay lean and wary, even in the face of success.

16

Technology and Literature

Technology is a form of communication. Because it is communication, technology both echoes in our literature and seamlessly continues human discourse into another domain that is wordless. Suppose you want to tell a friend how to go from Houston to Detroit. You might write out the sequence of roads and turns that would get her there. Or you might prepare a map. On the other hand, you might do something more abstract; you might tell her what it feels like to drive to Detroit—about the ride and the sights you see on the way.

The engineers in Detroit have another way to describe the trip. They design the machine we use to make the journey. They create the experience of the trip, give it its form and texture. Those engineers are using the automobile to tell you their own concept of what that experience should be. The feel of it, the sense of motion, the beauty of the auto, the way the car fits into your life and shapes it—these are all things the designer communicates in a remarkably compact and efficient way.

This fact was dramatically impressed on my wife and me the day we found a prefabricated desk that we needed for her computer. Since the box had been damaged by a forklift, the as-is price was next to nothing. It was a big, complicated, three-element item, with ten pages of assembly instructions. Putting it together was no job for the timid. We took the box home, opened it, and only then found that the instructions were gone. There lay thirty precut pieces of wood, hundreds of metal and plastic fittings, and no hint as to how they were supposed to fit together.

At first it was devastating. Then I realized that I could consult the designer directly! Why not just look at the parts and listen to the clear logic they represented? Why was this piece notched and drilled the way it was? Why did some fittings have little ribs while others did not? In the end, we had been relieved of the tedious and confusing intermediary of written instructions. When we worked from inside the designer's head, it was as if the designer were guiding our hands. The whole thing went together smoothly, and I came away with a real respect for this anonymous person whose essential sense of simplicity and elegance we had come to know.

We are perfectly happy to acknowledge other nonverbal forms of communication—pictures, music, body language. Technology is the largest such presence in our lives, and it speaks to us with a powerful clarity and directness. Sometimes it speaks of venality and greed. But good technology speaks of beauty and order. The most effective creators of our technologies are driven by the need to share the vision that has formed in their minds. The other side of this coin is that technology spills over into our writings all the time. Sometimes the spillover is obvious, as in Robert Pirsig's *Zen and the Art of Motorcycle Maintenance*.[1] Sometimes it is not. Few people are aware that Henry David Thoreau wrote "civil engineer" after his name, and that it was he who perfected the production of lead pencils in America.[2]

You see, in our highly professionalized and compartmentalized modern world we have artificially separated self-expression in writing and self-expression in making things with a line that cuts through our entire intellectual establishment. As I began the business of writing radio scripts, that division still seemed to be very clearly drawn. In the intervening years the line has become increasingly hard to find. Now I wonder how I ever imagined it existed in the first place. Of course, the division began to fade right from the beginning—in the programs upon which this chapter is based. As those scripts accrued, history itself began erasing the line. In 1988 I was still an engineer looking across the fence at a different world. When I look at creative self-expression today, I can no longer find that dividing fence at all, for history dramatically shows how little difference there is between inventing words and inventing machines.

Nowhere was that point made more clearly than it was when the Romantic poets sought to make sense of the Industrial Revolution. As the eighteenth century drew to a close, this group of radical intellectuals forcefully began to warn us that we must once more put ourselves into an equation that balances mind, machine, and nature. If you saw the movie *Chariots of Fire* back in 1981, you may well have missed the few lines of music that gave the movie its title. The title is a phrase from an English hymn sung at a funeral, and it occurs right at the beginning of the movie. The words, from William Blake's poem *Milton*, seem at first to portray the Industrial Revolution as a form of human evil. He says:

And did the Countenance Divine
Shine forth upon our clouded hills?
And was Jerusalem builded here
Among these dark Satanic mills?

Robert Burns said much the same thing when he first saw the fire and smoke of the Carron Iron Works in 1787. He reportedly fingered these lines on a grimy window pane:

We cam na here to view your warks
In hopes to be mair wise,
But only, lest we gang to Hell,
It may be nae surprise.

The Romantic poets wanted to tap into nature's wild forces. Nature had not looked so pretty when life was a struggle to create minimal physical well-being in a seemingly hostile world. Now the new factories were providing goods and implements by which people could live more amicable lives. But those works had also started blotting out the pastoral world of preindustrial England. As they did, poets and artists began to make rural life into something it had never quite been. Sir Walter Scott saw nature as beautiful, but it was primordial—a dark and formidable gothic presence: "Farewell to the forests and wild-hanging woods; / Farewell to the torrents and loud-pouring floods." Percy Shelley, a little younger and more the creature of the fully evolved Romantic movement, saw nature in more benign terms: "Sweet oracles of woods and dells, / And summer winds in sylvan cells."

Hellish mills *were*, nevertheless, replacing both Scott's and Shelley's visions of nature with their harsh brush strokes of fire and iron. Yet it was William Blake who also said: "Nature without man is barren." He saw that *we* are ultimately responsible for reclaiming nature. That Blake hymn text ends like this:

Bring me my bow of burning gold!
    Bring me my arrows of desire!
Bring me my spear! O clouds unfold!
    Bring me my Chariot of Fire!

I will not cease from mental fight;
    Nor shall my sword sleep in my hand
Till we have built Jerusalem
    In England's green and pleasant land.[3]

Blake, a sensible observer of the human lot, outlines *our* responsibility. We cannot shrink from the mental fight of building a world fit for habitation. When he asks for his bow, arrows, spear, and chariot of fire, he reaches for tools with which to build that world. Blake is arming for a mental fight. He realized that from now on, nature would shine through the fire and mills only if we had the wits to make it do so.

The most formidable dissection of the new man-made industrial world was undoubtedly the one given us by Mary Shelley. I will never forget my first meeting with the creature constructed by her character, Victor Frankenstein. I was thirteen years old the night I made the lonely mile-and-a-half trek back home from the movie theater where I had just watched *Frankenstein Meets the Wolfman*. It was a wintry night with a full moon flickering through naked branches whipping in the wind, and, I tell you, I was flat-out frightened to death.

What makes the Frankenstein story such a powerful part of our folklore? Why is it so much more than just one more movie plot to be seen and forgotten? Mary Godwin, soon to become the wife of Romantic poet Percy Shelley, wrote it in the summer of 1816 while she and Shelley and other members of Lord Byron's counterculture entourage were vacationing with Byron in Switzerland. Mary was the nineteen-year-old daughter of the noted feminist writer Mary Wollstonecraft and the anarchist theorist William Godwin. The group talked about creating a modern gothic novel and agreed that each would attempt to write one. Mary was the only one who really succeeded, with *Frankenstein, or the Modern Prometheus.*

For someone so young to hand us a creature of such disturbing force was astonishing. But it came out of a hotbed of intellectuals steeped in a rising concern over what science and industrialization were doing to the world. Mary Shelley's antihero, Victor Frankenstein, tells us early on,

"My reluctant steps led me to M. Krempe, professor of philosophy, an uncouth man, but deeply imbued in the secrets of his science." Under Krempe's instruction, Frankenstein's Faustian quest for knowledge takes him to the terrifying secret of life. What that secret is, we cannot be sure, although it is clearly akin to the new electrical science of Franklin, Volta, Davy, and Faraday. His product, the monster, is more articulate, more intelligent, and more able to feel pain than his human maker. The monster produced by Frankenstein's intelligence and creative drive had his creator's intelligence and sensibilities, but in a grotesque parody. Frankenstein and his monster merge into one.[4]

Mary Shelley was unmistakably talking about the science-based technology of her day. The subject interested her. Later in her life she wrote biographies of famous scientists for the Reverend Dionysius Lardner's *Cabinet Cyclopaedia*. Her *Frankenstein* expressed her recognition of the dangers that lay in our new powers. She summoned up a monster that can be found in any of us—the monster that Victor Frankenstein released when he let himself be obsessed by technical knowledge. In the end, the monster portrayed his obsessiveness. Scientists are usually pretty sane people. Shelley's story reminds us what damage we do if we drop the discipline that keeps us sane, keeps our work rooted in the joy of simple curiosity—keeps the monster at bay.

That lingering image of obsessive science did not begin with Shelley, original as her monster may have been. Think of a scene you know very well: The laboratory is down the stairs, out of the light, equipped with bubbling liquids, strangely shaped glassware, and arcane electro-mechanical machines. The scientist is lonely, naive, and egomaniacal. He tells others about his humanitarian aims, while something in his animal nature drives him to darker things. This picture of the mad scientist is too strong and too persistent. We cannot shrug him off as just another piece of popular fiction. He is the image of science and technology that we revert to when we let down our guard. We have to ask where he comes from and what he means.

That scientist almost surely comes to us from the Faust story. The real Faust was a shadowy figure in early-sixteenth-century Germany—a kind of self-styled magician and hell-raiser. One place he shows up is in the records of the city of Ingolstadt, which was the same town Mary Shelley used as the home of Victor Frankenstein. Storytellers took up the legend of this character and recast him in the language of the Protestant Reformation. The Faust we know, the Faust who sold his soul for knowledge, was given his present form in 1607 by Christopher Marlowe in his book *The Tragicall History of Dr. Faustus*. At the same time modern science was taking shape as the companion of technology (Chapter 5).

Two hundred years later Mary Shelley took Faust a step further. Knowledge alone was not enough for Victor Frankenstein. He had to *create* life as well as *understand* it. The evil force that served Faust was alchemy. Frankenstein rode electrochemistry on his trip to hell. Later in the nineteenth century, Faustian mad scientists added hypnotism and then the mysterious new forces of radiation—X rays and radium. The late nineteenth and early twentieth centuries replaced Faustian scientists with Faustian technologists such as Captain Nemo and Fritz Lang's cinematic soul-eating mechanical city of *Metropolis*. Naturally, today's mad scientist is wed to a computer and works in a biological laboratory.

Each new scientific or technological discovery calls forth new fears that we will be unable to control it. Each new discovery brings out a Faust or a Frankenstein or a mad scientist in some new incarnation. Robert Louis Stevenson added a telling twist. The monster, Mr. Hyde, emerged from gentle Dr. Jekyll when Jekyll lost control of his knowledge. Jekyll and Hyde touches something about our nature that we all understand instinctively. The Faustian mad scientist is fictional shorthand for describing our potential lack of control, not so much the control of our science-based technology as control of ourselves. The mad scientist is really a facet of our own human nature, and the literature surrounding him is no more than wariness of our own potential for evil.

Frankenstein's creature, that human attempt to create human life, reemerged in yet another form not very long after Mary Shelley told us her version of him. This time the beast was a human attempt to re-create human thought. It was Charles Babbage's attempt to build the first programmable computer, or analytical engine. Babbage knew the late Romantic poets and writers. He read their work and exchanged letters with them. I have little doubt that he drew upon the same forces that fueled their vision. That's why his friendship with Ada Byron is so unsurprising.

Just after Lord Byron's daughter, Ada, was born in 1815, Byron wrote: "Is thy face like thy mother's, my fair child! / Ada! Sole daughter of my house and heart?" Actually, Lord and Lady Byron were separated even then, and he made no claim on Ada's custody. Ada was timid and sickly as a child, but she had a sharp, analytical mind. She wanted to be a mathematician. No doubt that desire came from the tough and manipulative Lady Byron, who also loved mathematics. In fact, Byron had ridiculed Lady Byron for her passion, calling her the "princess of parallelograms."

When Ada was seventeen, she met Babbage on a visit to London. Babbage was then forty-one years old and a leading mathematician. Despite the age difference, a kind of mental chemistry formed between them. Some years later, after she was married to Lord Lovelace, she continued her mathematical studies under Babbage's instruction.

By this time, Babbage had begun work on the analytical engine. It was a brilliant mechanical invention, with all the basic elements of a modern computer. Babbage eventually ran aground on problems connected with manufacturing the physical machine. He never completely finished it, but he was highly successful in setting up the working principles that later computer designers would follow. He was a century ahead of his time.

He asked Ada to write a description of the machine's operation and capabilities. Her work was published in *Taylor's Scientific Memoirs* in 1843 as a series of seven notes. She was just twenty-seven at the time, but the notes display a late-twentieth-century understanding of what the computer is and what it can do. In the most famous quotation from the notes, she says:

> The analytical Engine has no pretensions whatever to originate anything. It can do whatever we know how to order it to [do]. It can follow analysis; but it has no power of anticipating any...truths. Its province is to assist us in making available what we're already acquainted with.

Ada died of cervical cancer only a few years later. In 1980, on the anniversary of her 165th birthday, the Defense Department announced a powerful new computer language, and they named it Ada in her honor. Most computer language names (with the notable exception of Pascal) are acronyms, but not this one. It was a simple tribute to a rare person who saw the future with remarkable clarity.

As a footnote to the Babbage-Ada connection, it is worth noting that the year after Ada Byron published her notes on the analytical engine, Elizabeth Barrett wrote a letter to Robert Browning. Someone had criticized Tennyson's poetry, and Barrett was annoyed. She said,

> That such a poet should submit blindly to the suggestions of his critics is…as if Babbage were to take my opinion & undo his calculating machine by it.[5]

So these people knew their works were all of a piece. They talked with one another, and each drew upon what the others were doing.

This was not just a phenomenon of the nineteenth century. I could pick from any number of twentieth-century examples, but I shall settle upon my favorite. You may have read the book *On the Beach*. It was a wrenching, understated story told through the eyes of the last survivors of nuclear war, waiting to die in Australia. In 1960 this best-selling book, and a fine movie based upon it, made a disturbingly realistic case against the nuclear stalemate.

The author was Nevil Shute Norway, who was born in England in 1899. After serving in World War I, he trained at Oxford to be an aeronautical engineer. From 1922 to 1933 he worked in airplane and dirigible companies. He also published his first four novels, one of which was made into a movie. He did not want the engineers he worked with to think he was frivolous, so he wrote under his Christian names, Nevil Shute. His book writing slowed in the latter 1930s as he threw his energies into forming an airplane manufacturing company called Airspeed Ltd. The company built airplanes for Douglas, Fokker, and others.

By 1938 he had built the company up to a thousand employees, but he was then eased out of its management. In his autobiography Norway is philosophical about that part of his life. He says it was probably time to move on because two kinds of people shape companies, *starters* and *runners*. He had creatively started the company, but he was not suited to running it day by day. So he left the company and turned back to writing. He wrote roughly one book a year until he died in Australia in 1960. Today we know Nevil Shute for many books and movies, such as *On the Beach* and *The Pied Piper*, that were made from them.

Shute's books are low-key, but his plots are assembled like Swiss watches: Every piece fits perfectly, and you simply cannot put one down after you are fifty pages into it. They also contain astounding technical realism—far more than you might think could hold his readers' attention, let alone keep them spellbound. His typical protagonist is reserved, capable, and usually somewhat mousy. Some of his early work (*An Old Captivity* and *In the Wet*, for example) flirted with strange notions of mysticism. It was a little like John Steinbeck's early work in that regard. But Shute always came back to powerful storytelling. *The Legacy* was Shute in top form. It was also made into a movie, but you are more likely to remember the television miniseries based on it, renamed *A Town Like Alice*.

By the way, *On the Beach* was not at all typical. Shute was no writer of Greek tragedy. The engineer in him said that we can solve our problems and must not let them beat us. Despite his extraordinary success— twenty-three books, largely best-sellers—he was first of all an engineer. He once said that he thought of himself not as an author but as an engineer who wrote books. His last book, *Trustee from the Toolroom*, told of a middle-aged engineer who was obliged to do an immensely complicated bit of smuggling. The book was a runaway fiction best-seller in 1961, yet it could have doubled as an engineering text. If you have never read any Shute, try him; you will be surprised.

To my mind, Shute's most remarkable book was *No Highway*. To understand its near prescience you need to know the state of air transport when it was published, in 1948. The redoubtable Boeing 707 jetliner went into service ten years later, in 1958. But the English got a big jump on Boeing with their de Havilland Comet. The Comet began to carry passengers six years before the Boeing 707, in 1952, and four years after *No Highway* came out.

Disaster struck the Comet after one year of service. A Comet, flying out of Calcutta, disintegrated in a thunderstorm. When investigators could find no other cause, they blamed it on the storm. Eight months later a second Comet blew up in clear sky, twenty-seven thousand feet over the island of Elba, off the coast of Italy. It was hard to recover much from the ocean, so that crash went undiagnosed. When a third Comet exploded over the Mediterranean three months later, the whole fleet was grounded.

An intensive search finally yielded some wreckage, which showed that the failure had occurred in the cabin area. So engineers did a full-scale fatigue test of an actual airplane. They varied the cabin pressure hydraulically while they flexed the wings. After three thousand pulsations, a crack appeared near a cabin window and quickly spread. It turned out that the Comet's designers had overlooked stress concentrations at rivet holes near the windows. The windows were redesigned and a new, safe Comet went into service in 1958, still five months ahead of the Boeing 707.

Shute's company had worked for de Havilland, and *No Highway* was published while the Comet was in its final design stages. The book is about a new airplane called the Reindeer that has mysteriously crashed in Canada. (Our ears prick up; along with *Dasher, Prancer,* and so forth, *Comet* was also one of Santa's reindeer.) An engineer, a structural theoretician named Theodore Honey, is sent to investigate the crash. Honey has his own theory, that Reindeers are overwhelmingly likely to suffer a fatigue failure after about fourteen hundred hours in the air, but no one takes him seriously. Halfway across the Atlantic Ocean, Honey, who is pretty oblivious to his surroundings, discovers that he is riding in a Reindeer.

A few questions reveal that this particular plane has been in service just about fourteen hundred hours. Honey suddenly has to assume responsibility for saving two hundred people who feel no need of being saved from anything. (If you choose not to read the book, I leave you to find a video of the movie, titled *No Highway in the Sky*, to see what Honey did. It stars Jimmy Stewart as Honey, with Marlene Dietrich.)[6]

How did Shute anticipate the Reindeer disaster? Henry Petroski offers a theory that, while not dramatic, is convincing. He thinks Shute followed his engineering instinct, which was very good, and it simply took him where real life was destined to take the Comet. Perhaps we can word that in another way. By now, we see something that innumerable writers and engineers have also seen. It is that technology is itself a form of storytelling. Tell the story, and the story will complete the technology. Build the machine, and the machine will tell its story. Just as my computer cabinet put itself together without an instruction manual, Shute unknowingly predicted the destiny of the Comet by following the machine itself. What Mary Shelley really did was to help us understand modern machinery by simply letting her mind drift about the questions of modern machinery while she wrote a gothic horror story. In the end, the machine itself is a story.

17

Being There

We come at last to the forbidden first person, the *I am*. No story is right until the teller is part of it. Yet a peculiar mischief is abroad in the land of science and engineering. It is a mischief born out of the noblest of intentions. For decades it has spread like the flu, far beyond the technical journals that gave it birth. The intention is to let us stand like blindfolded Justice—pure, objective, and aloof. To do this, we write about our work without ever speaking in the first person. We try to let fact speak for itself. Instead of saying, "I solved the equation and got $y = \log x$," we write, "The solution of the equation is $y = \log x$."

We turn our actions into facts that are untouched by human hands. To some extent we must do that. Our facts should be sufficiently solid that we do not need to prop them up with our desires. Third-person detachment has its place, but my own person is not so easy to erase.

Suppose I think another engineer, whom I shall call Hoople, is wrong. I am not objective about Hoople, but I must appear to be. So I write, "It is believed that Hoople is incorrect." That's a cheap shot. I express my thoughts without taking responsibility for them. I seem to be reporting general disapproval of Hoople. In the unholy name of objectivity, I make it sound as though the whole profession thinks that Hoople is a fool. Now radio and TV journalists are doing it. I cringe every time I hear, "It is expected that Congress will pass the bill." Who expects that? The announcer? The Democrats? A government official? Maybe the soy sauce lobby is the expectant source. So instead of objectivity we get obfuscation. If our work really occurred in objective isolation, we could write about it that way. But people are present. They think and they act. If we fail to represent human intervention accurately, we are dishonest, and objectivity becomes meaningless.

The things we make tell the world what we are. Real objectivity means admitting our actions and our thoughts as long as they are part of the story. Instead, the language of technology and commerce is filled with things like this: "A new reactor is described. Its features are thought to be superior to those of the Hoople reactor." I have not eliminated my id here. I have merely hidden it in a box, and there it has grown large. If I really spoke objectively, I would say, "I've tried to improve Hoople's reactor. I'll describe my design. Then you can judge it." These are much more than questions of style. They are matters of honesty and directness. "If you're ashamed of it, don't do it," goes the old saying. In practice, that means doing things we can talk about directly and using language that admits what we have done. When we speak in language that is clean and translucent, then objectivity takes care of itself.

So my radio program about technology and creativity must be, in some measure, autobiographical. For me, a technologist, to write about what I do without speaking in the first person is like trying to telegraph my dislike of Hoople without admitting it. Yet for me to assume the first-person voice was difficult. Like the first time I attempted a back dive off a diving board, it violated a lifetime of training in self-protection. That back dive opened the door to all kinds of dangers—the danger of exposure, of excess, and of losing academic plausibility.

So I conclude this exploration of technology and attitudes by recalling that particular backward leap into space—by consciously inserting myself into the story. That is not to say that my life has been a sequence of great moments in history. I did not stand on the deck of the sinking *Titanic,* nor did I close the switch that opened the sluice gates of Boulder Dam. Shakespeare's Henry V tells his troops before the Battle of Agincourt that "gentlemen in England now a-bed shall think themselves accursed they were not here...upon St. Crispin's Day."

I have no particular sense that I was there upon technology's St. Crispin's Day. Yet, like everyone, I have been a constant first-person witness to, and participant in, the technology and creativity that has unfolded in all our lives. The radio program soon demanded that I display all that, not just as it has unfolded in the pages of books, but as it has unfolded in my own experience as well. I began dealing with technology from inside my own skin. When, for example, I got around to telling the story of the telephone, I began by talking about my father, who was raised in the little Swiss-American community of Nauvoo, Illinois.

He told me about coming home from school one day in the late 1890s to find his mother shouting into a strange box mounted on the wall. It was the first telephone he had ever seen, and I have often wondered why she bought it back then. What had she seen in this strange new gadget? Bell received his telephone patent in 1876, and by 1880 only one American in a thousand had a phone. When my father came home that day, the number was still only about one in seventy. In their first quarter century telephones did not by any means sweep the country.

When we go back and look through old telephone advertisements, we see what had to happen for this novelty to become what it is today. The telephone was first seen as a replacement for the telegraph. Advertisers pointed out that telephones were better than the telegraph for transmitting news, ordering groceries, and sending urgent messages. Brevity had been primarily important in using the telegraph, and that attitude carried over to the telephone. Gossip and idle chat were discouraged. Telephone companies complained about the frivolous use of telephones and told their users to be businesslike. Their machines were, after all, important.

Not until the 1920s did telephone companies catch on to what people really wanted from this wonderful machine. People wanted to be drawn into a kind of living tether with each other. The Bell Company started telling long-distance customers, "Your voice is you!" In the 1930s AT&T first suggested that we "reach out and touch someone." Today, even in business, that is how we use telephones. Telephones unite our scattered families and keep friendships alive.[1]

Alexander Graham Bell himself predicted the social use of the telephone, but early makers and users failed to see it the same way. It used to bother me that, up to the day he died, my father was never able to relax and chat with me on a long-distance telephone call. It took the next generation to see that the inherent use of the telephone was social. Our tools teach us. They drive our minds and evolve their own roles in our lives. Some do it more quickly than others. It took a long time for the telephone to explain itself to us.

Just as my understanding of the telephone was shaped by my father's experience of it, my understanding of the typewriter was likewise shaped, in a way no book could have done, by an odd artifact on my study shelf. It is my great-grandfather's letter book, a bound volume with only tissue-paper pages remaining. He wrote letters with a dipped ink pen on regular paper that had once lain between the tissues. Before he tore the letters out and mailed them, he blotted them on the tissue, leaving copies behind. The fuzzy imprint of one brief, telegraphic letter, written in 1891, says: "Gentlemen, Please quote me the expense of shipping car load of saw dust from your station."

Why did he use such a crude method? Carbon paper was an invention that had been around for twenty-two years in 1891, but you couldn't press a dipped pen hard enough to use it. With a typewriter and carbon paper, copying would have been a cinch, and the typewriter was around by then too. So let us look for the reason why it had not yet found its way to Great-grandpa.

A man named Christopher Sholes started to develop a workable typewriter in 1867. He drew in co-inventors and together they made improvements. In 1872 they found a manufacturer in the small-arms maker Remington. When the Civil War ended, Remington needed a peacetime product. The first Remington machine came out in 1874, and by 1878 it had developed into something very much like the modern manual typewriter. So why was Great-grandpa still using ink pens and blotter copies in 1891?

Only five thousand typewriters had been sold by 1880. Typewriters were an exciting novelty—people loved to watch demonstrations of the machines in stores—but they were not something people saw any use for in their daily lives. Letter writing was a Victorian rite, something done according to rules and conventions. Good handwritten letters were the mark of a lady or a gentleman. A typed letter looked like a printed flyer. Even a backwoodsman was offended when he got his first typewritten letter. He replied, "You don't need to print no letters fer me. I kin read writin'."

Remington did badly at first because he tried to sell these machines to household buyers. When he finally triumphed, it was because the business world saw the value of the machine. Ten times as many typewriters were sold between 1880 and 1886 as were sold in the preceding six years. When Great-grandpa wrote his letter to the lumber company in 1891, the times were just catching up with him. By then, most small businesses had at least one typewriter.[2]

Like so many other machines, the typewriter had to teach us what it could do after it was invented. After all, we still like to receive a hand-written note from a friend; but how long could we survive in today's business world using handwritten letters and blotter copybooks?

Now, of course, I write everything on a word processor and (often as not) send it off by E-mail without ever touching paper. In my early days on a word processor, throughout the 1980s, I still had to put all my mail on paper—printed by a dot-matrix printer. For a printer stand, I used my mother's old floor-model sewing machine, made in 1905. The sewing mechanism folded down, leaving the top flat. That's where I put the printer. The paper sat on the treadle below, and it fed up into the back of the printer. It was all so neat you would have thought the sewing machine had been *designed* as a printer stand.

My mother never gave the machine up for an electric model. She liked its movement, the hand-foot coordination. I like the florid art nouveau design of the cast-iron stand, the grain of the walnut top, and the pretty wooden molding. I like knowing that, if push came to shove, I could actually sew with it.

The invention of sewing machines was brought on by the Industrial Revolution. Suddenly so much fabric was being produced that someone had to invent a machine to sew it all. In 1790 an Englishman named Thomas Saint patented the crude forebear of today's machines. For the next fifty years, patent after patent chipped away at the problem of making a machine do the complicated things a human hand does when it sews. The strongest all-around patent was one filed by Elias Howe in 1846. It led to a spate of thinly veiled copies and a patent war. The major inventors finally had to form a sewing machine trust that paid Howe a handsome royalty. The industrial giant that emerged from this trust was the Singer Sewing Machine Company.

My mother's sewing machine was made by the Willcox-Gibbs Company, which was founded around James Gibbs' patent for a chain-stitching machine in 1856. The company was one of many that competed with Singer by making less expensive machines. It stayed in business at least through the 1960s. In 1859 *Scientific American* magazine wrote this about the Willcox-Gibbs machines: "It is astonishing how, in a few years, the sewing machine has made such strides in popular favor [going from] a mechanical wonder [to] a household necessity."[3] That's exactly what happened. Sewing machines took the country by storm. They were revolutionary. They changed American life.

When I was six, my mother had me lie down on a piece of butcher

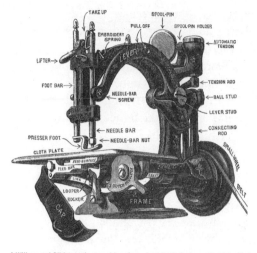

A Willcox and Gibbs sewing machine, from *Appleton's Cyclopaedia of Applied Mechanics*, 1892.

paper. She drew a line around me and used it as a pattern. She sewed a human figure, stuffed it with cotton, hem-stitched a face on it, decked it with hair of brown yarn, and clad it in a suit of home-made clothes. Then she gave me this life-sized alter ego, this doppelgänger, as a playmate.

I look at the old machine and see my mother's quirky imagination, her care for me, her highly honed mechanical skills. I remember American home life as it was so powerfully affected by these beautiful and complex old engines of our ingenuity.

Now, if my father, my great-grandfather, and my mother informed my understanding of the telephone, the typewriter, and the sewing machine, my understanding of radio was far more direct. For I was in grade school in the 1930s. I was a member of the first generation to be raised with the radio as a constant companion. My family would gather around the radio in the evening. It was just a simple wooden box with a rounded top. But I would look in the back and see its bright glowing tubes—a kind of miniature Christmas tree with the mysterious power to pluck adventure out of the sky and drop it into our living room. After school we listened to *The Lone Ranger* and *Jack Armstrong*.

But in the evening we had to listen to the grown-up stuff: Jack Benny, Fred Allen, all kinds of good music. One October evening my father tuned in to *The Mercury Theater*. He caught it just a minute or two late. An agitated network announcer was telling about a huge meteor that had landed in central New Jersey. Then he switched to a mobile unit. Something strange was happening. Something was coming out of the crater, something alive!

The game was afoot. This was, of course, the most famous radio show ever broadcast—Orson Welles' version of H. G. Wells' story *The War of the Worlds*. By 7:15 P.M., the full fury of the invaders from Mars was clear. They were moving across America toward my gentle home in Minnesota, killing everyone and everything in sight. When the station break came, I laid my delicious thrill of terror before my father. "Was this real?" I asked. He smiled and said, "I don't know. We'd better stay tuned." I was almost disappointed when it turned out to be make-believe. Earth was safe, at least until World War II.

Radio was what the Grimms' fairy tales had been for children before me. It stretched my mind. It showed me the world of good and evil, honor and deceit, pain and pleasure. It sketched the story and let my mind fill in the details. I heard Joe Louis—larger than life—destroy the German champion, Max Schmelling, two minutes into the first round. In 1937 I heard the announcer when the zeppelin *Hindenburg* suddenly caught fire. His voice rose over the crackle of static and flames while he watched what no human being should ever have to watch. I learned the difference between fantasy and reality when he broke down and wept, "Oh, the humanity!"

There is too little of that on radio today. Commercial radio is for commercial messages. The trick is to fragment our attention span—whatever falls between commercials has to be simple. The radio of my childhood was more naive. It copied the stage, the theater, and the town meeting. It engaged us, it led us into our right brain, it touched our hearts, and it made us free.

If you too are old, you remember how you grooved on stories and daydreams when you were nine years old in the 1930s! Radio provided stories and daydreams. Radio gave you the words, and your mind drew the pictures; that was a powerful combination. It was rare to have more than one radio in a house in those days, and you listened to it when your parents said you could. But listening always ended too soon. You were always called away—to school, to supper, to chores, and to bed.

There was a way around the problem, but it was not easy. You could build your own crystal set. That was a simple, primitive radio whose heart was a polycrystalline lump of galena set in lead. The crystal worked as a rectifier, in place of a radio tube. You pecked away at its surface with a fine wire probe called a cat's whisker. Sooner or later you hit just the right facet of the crystal—one that responded to the station you had set on a homemade coil. The signal was weak; there was no amplifier. You listened to it with earphones. But with your own crystal set, you would be able to pull the covers over your head and listen to your heart's content after your mother had said good night. You would be able to listen to music or *I Love a Mystery,* and no one would know.

Mechanics Illustrated or any of a hundred how-to books explained, in formidable detail, how to make a crystal set from hardware-store parts. But you were nine years old, and something always went wrong—a loose wire, a badly wound coil. So you never quite figured out how to make your own radio to play under the bedcovers. In the end you were left to read comic books with a flashlight.

The radio permeated American life with amazing speed after it was invented late in the nineteenth century. Just a few years later, the 1910 Boy Scout manual told us how to earn a merit badge in radio. You had to draw the complete circuit diagram for a receiver set from memory. You also had to build your own set using a tube, not a crystal. Then you had to pick up a signal twenty-five miles away from a transmitter.

The problem is, that was for fourteen-year-olds. But when you were only nine, making a radio was far easier to dream about than to do. So you dreamed and wished. Of course, that wish actually came true. Today we can all let Chopin and Britten wash us into slumber. But you still wish that, just once, you could have teased one of those temperamental crystals into singing you to sleep.

During the 1920s—before TV, talking movies, or decent recordings—my mother ran her own radio program. A sheet-music store hired her to play the piano and sing the latest songs so people would buy the music. She started her brother in radio by letting him sing "On the Road to Mandalay." That did not lead to musical stardom, but he went on to become a noted broadcast journalist in the 1930s and '40s. My father was a newspaperman and editor through the 1920s, '30s, and '40s. I was, in short, a child of the media—raised to understand that clear, honest, entertaining communication was a virtue on roughly the same level as eschewing theft and murder.

But the family trade was not for me. I suffered from what today is swept under the confusing label of *dyslexia*. I could neither read nor write acceptably, and I was beleaguered by a severe stammer. I looked for fulfillment not on any public stage but in the workshop. I built model airplanes, tree houses, many of my own toys, and anything else I could think of.

My parents seemed content with this alternative, although I seriously challenged their patience one spring afternoon in the waning days of World War II. The Japanese, in their eleventh-hour desperation, had started releasing balloons into the westerlies that blow across the northern United States (see Chapter 10). Each one carried a small incendiary bomb. Luckily, America was less flammable than the Japanese had hoped. This first assault on the American mainland did little damage, but it created enormous media interest.

So I set about to launch my latest construction. It was a six-foot-tall hot air balloon made of white tissue paper and painted with my concept of Japanese characters. I had built a small oil-fired heater to buoy it into the sky. I was puzzled and hurt when my father saw what I was doing and told me to quit it right away. A media child I may have been, but I had not yet learned that neither media people nor their children were allowed to manufacture news.

All that came back to me in a rush in the last days of the former Soviet Union. I was chatting with a visiting Russian scientist when he blandly asserted that the American media were all under government control. What could I say to him? He had no way of understanding the burden of responsibility that vests in people who enjoy genuine freedom of expression. I had no ready way to explain the power of the values that lay at the center of my upbringing.

And so I grew up, finished high school, went through college, did a master's degree, and was finally drafted into the army in the waning days of the Korean War. The year 1954 found me with the Signal Corps at Fort Monmouth, New Jersey. I was designing test apparatus for materials that could be made into a radical new device called a transistor. None of us had a clue as to how we were helping to change life on planet Earth with that new gadget. I simply worked away, washing dishes, standing in formations, and inventing cutting-edge high-temperature furnaces.

One hot summer's day I had a day off, so I left Fort Monmouth, walked to the highway and put out my thumb. In my writing case was a thermodynamics text. I meant to find a quiet place to write letters and study Einstein's theory of specific heats. The car that picked me up was going to Princeton. That sounded good. I asked the driver, "Isn't that where Einstein lives?" He allowed it was. I got out at the university, asked where Einstein was, and was told he worked two miles outside of town at the Institute for Advanced Studies.

So I walked to the institute and sat for a long time in the commons room, studying thermodynamics and watching very smart people coming and going. But no Einstein. At length I gave up my ad hoc pilgrimage and started back to town. Where the road turned, I looked over my shoulder and saw a figure two blocks back. The sun behind him cast a brilliant halo through a mop of frizzy white hair. I stalled, watching a golf game, while Einstein himself passed me and strode on into town. I fell in behind him.

He walked vigorously, greeting friends and neighbors. Then he stopped and laid his briefcase on a hedge. I was terrified. Did he know I was following him? I don't think so. He was just taking off his blue sweater. As I passed him I saw suspenders over a T-shirt holding baggy trousers. He wore sandals—no socks. That much fit the stereotype. What *didn't* fit was his substance. He was then seventy-five, with less than a year to live. But he had an earthy muscularity. He had physical grace, strength, and coordination. How many people know that he was a good violinist? Einstein was more than airy energy and light. He had mass and physical presence as well.

My now-battered thermo text sits on my shelf without Einstein's autograph in it.[4] I was far too shy, unformed, and uncertain to speak to him. Three years later, by a fluky coincidence, I designed an apparatus for his son, the civil engineer Hans Albert Einstein. I didn't mention Princeton to him. Six years afterward, I met his granddaughter. Didn't mention it to her either.

A lifetime later, the other famous names of that epoch—Eisenhower, Chiang Kai-shek, Churchill—all fade against the light, energy, and mass of that quiet man, standing by a hedge, juggling stars and forces and fields in his head—that man who made us understand that the world is more than it seems to be. I went on to learn about Einstein's statistical mechanics, his theory of Brownian movement, Bose-Einstein condensation. I learned about his philosophical ambiguities, contradictions, and complexities. But overwritten on all that is, for me, a neighbor greeting other neighbors and smelling the roses. My Einstein will always be that scruffy old man on a summer's day in Princeton—that gentle man who surely *would* have autographed my thermodynamics book, had I only asked him to.

I said at the outset that I was not at Agincourt on St. Crispin's Day. I was not at mission control when we walked on the moon. My life has been no more remarkable than yours, perhaps less so. I was not at Trinity when we triggered the first atom bomb (thank God!), yet I first studied thermodynamics from that battered book with the same person who invented the means for detonating the "Fat Boy" bomb. I did not invent the integrated circuit, but I was one of those who nursed the transistor in its cradle. I never met Einstein, but I have known his presence. I have operated a surveyor's transit almost identical to the one

George Washington used, and I have driven cars on the roads I once helped to lay out. I have designed, analyzed, made working drawings, created theories for phenomena, and seen the functioning machines that have been the ultimate result of that work.

Run that inventory on your own life. I have been there, but so have you. You and I are part of all this. You and I have made the world we live in. Let us, you and I, lay the passive voice aside for just a little while and send up our own joyful cry of "I am!"

Be in their flowing cups freshly remember'd,
This story shall the good man teach his son;
And Crispin Crispian shall ne'er go by,
But we in it shall be remembered;
We few, we happy few, we band of brothers;

—William Shakespeare, *King Henry* V,
act 4, scene 3

Correlation of the Text with the Radio Program

People interested in the radio program upon which this book is based might wish to know which radio episodes have been developed into each chapter. (Those episodes, almost all from the first year's broadcasts, and all that have followed them, may be found on the *Engines of Our Ingenuity* Web site, http://www.uh.edu/engines.)

When I write XX/YYYY for a number, that means that the old episode (XX) was subsequently rewritten and rebroadcast at a much later date (but prior to the year 2000) under the latter number (YYYY). I have used a very few new episodes from beyond the first year of the program. Those are ones with numbers higher than 195.

1. MIRRORED BY OUR MACHINES
12, 16/1410, 20/540, 52/1363, 55/1381, 137/1559, 68/1407, 57/1384, 55/1381, 10/1328, 18/1413, 89, 96/1480, 88/1518, 173

2. GOD, THE MASTER CRAFTSMAN
3/1142, 9/1311, 107, 33, 99/1469, 46/1294, 97/1530, 123, 144, 131

3. LOOKING INSIDE THE INVENTIVE MIND
194, 190, 4/1165, 73/1475, 74, 90/1484, 119, 174, 142/1546, 143, 162, 23/1323, 84/1510, 152, 148

4. THE COMMON PLACE
247, 40/1347, 163, 38/1282, 135, 165, 48/1295, 78/1468, 157, 170, 130

5. SCIENCE MARRIES INTO THE FAMILY
299, 166, 113, 5/1307, 183, 76/1508, 147, 192, 98/1461, 121

6. REVOLUTION
122, 105, 168, 200, 69/1440, 82/1509, 109

7. INVENTING AMERICA
161, 43/1350, 22/1317, 28/1085, 141, 60/1385, 30/1365, 128, 1/264, 36/1420, 8/1373, 156, 181, 159

8. TAKING FLIGHT
328, 153, 39/1351, 118/1545, 188, 184, 85/1489, 114

9. ATTITUDES AND TECHNOLOGICAL CHANGE
1156, 26/1318, 72/1506, 116/1543, 58/1390, 13/1180, 185, 31/1338, 193, 81/1473

10. WAR AND OTHER WAYS TO KILL PEOPLE
1160, 1091, 35/1418, 115, 7/1369, 108, 134, 91/1525, 42/1364, 71/1448, 179

11. MAJOR LANDMARKS
24/1254, 29/1359, 34/1309, 124, 45/1300, 54/1380, 2/1125, 27/1319

12. SYSTEMS, DESIGN, AND PRODUCTION
63/1434, 94/1482, 92/1470, 101/1252, 133, 145, 199

13. HEROIC MATERIALISM
17/1405, 19/1158, 59/1425, 189, 61/1426, 87/1488, 191, 197, 25/1361

14. WHO GOT THERE FIRST
1098, 70/1443, 6/1084, 14/1397, 15/1393, 11/1330, 125, 64/1435, 160/705, 50/1372, 32/1342, 37/1362, 86/1575

15. EVER-PRESENT DANGERS
1066, 126, 140, 62/1429, 106, 67/1438, 47/1332, 132

16 TECHNOLOGY AND LITERATURE
187, 18/1413, 41/1337, 129, 102, 110, 112

17 BEING THERE
431, 93/1487, 100/1532, 167, 195, 176, 178, 56/1423

Notes

CHAPTER I

1 The sequence leading to the invention of agriculture is described in: W. Stevens, "Dry Climate May Have Forced Invention of Agriculture," *New York Times*, April 2, 1991, section B. (based on work by Frank Hole and Joy McCorriston). Jacob Bronowski provides a cogent discussion of the mutations of grain-yielding grasses in *The Ascent of Man* (Boston: Little, Brown and Co., 1973), Chapter 2.

2 Here is a brief list of additional approximate human equivalents for units of measurement beyond those mentioned in the text. You will surely think of others as you scan this list.

> inch: thumb digit
> foot: human foot
> yard or meter: distance from nose to tip of an outstretched arm
> quart: most liquid we could drink at one sitting
> decibel: least change in sound level we can hear
> BTU, kcal, or Joule: least energy we can sense in our coffee cup
> year, month, or day: time needed for natural experiences to recur
> month: menses
> second: order of magnitude of human reaction time
> mile per hour: speed of slow human locomotion

I treat the matter of dimensional similarity in my text, *A Heat Transfer Textbook*, 2nd ed. (Englewood Cliffs, N. J.: Prentice-Hall, Inc., 1987), Chapter 4.

3 A good account of George Caley's work is given in: P. Scott, *The Shoulders of Giants* (Reading, Mass.: Addison-Wesley Publishing Co., 1995). See especially Chapters 2 and 3.

4 For the story of the Samurai sword and ceremony, see J. Bronowski, *The Ascent of Man*, Chapter 4, "The Hidden Structure." This is also available on videotape and film. For more on Japanese swords see K. Sato, *The Japanese Sword*, trans. by Joe Earl (Tokyo: Kodansha International Ltd. and Shibundo, 1983).

5 W. Stevens, "The Man with the Blue Guitar," *The Collected Poems of Wallace Stevens* (New York: Albert A. Knopf, 1982).

6 L. White Jr., *Medieval Religion and Technology* (Berkeley: University of California Press, 1978), 105-120.

7 More on the streamlining movement can be found in an exhibit review: C. K.

Hyde, "Streamlining America: An Exhibit at the Henry Ford Museum, Dearborn, Michigan," *Technology and Culture*, 29, 1 (1988): 123–129.

CHAPTER 2

1   Kenneth Clark tells of the construction of Gothic cathedrals in general and of Chartres Cathedral in particular in *Civilisation: A Personal View* (New York: Harper & Row, 1969).

2   The story of Eilmer and his glider flight is told by L. White Jr., *Medieval Religion and Technology* (Berkeley: University of California Press, 1978).

3   For more on cathedrals, Cistercians, and perpetual motion, see J. Gimpel, *The Medieval Machine* (New York: Penguin Books, 1976). The best full story of perpetual motion is found in A. W. J. G. Ord-Hume, *Perpetual Motion: The History of an Obsession* (London: George Allen & Unwin Ltd., 1977).

4   For the state of the mechanical clock at the end of the medieval era, see K. Maurice and O. Mayr, *The Clockwork Universe: German Clocks and Automata* (Washington, D.C., and New York: The Smithsonian Institution and Neal Watson Academic Publishers, 1980). And for the history of the hourglass, see R. T. Balmer, "The Operation of Sand Clocks and Their Medieval Development" *Technology and Culture*, 19, 4 (1978): 625–32.

5   Much has been written on the fourteenth- and fifteenth-century plague. See, for example, R. S. Gottfried, *The Black Death: Natural and Human Disaster in Medieval Europe* (New York: The Free Press, 1983).

6   Paul Nahin tells Kelvin's role in the aging-of-the-Earth controversy in "Kelvin's Cooling Sphere: Heat Transfer Theory in the 19th Century Debate over the Age-of-the-Earth," in E. T. Layton and J. H. Lienhard, eds., *History of Heat Transfer: Essays in Honor of the 50th Anniversary of the ASME Heat Transfer Division* (New York: ASME, 1988), 65–85.

7   For Fourier's theory you may go to the horse's mouth: J. Fourier, *The Analytical Theory of Heat* (New York: Dover Publications, Inc., 1955). I deal with the so-called semi-infinite region calculation that Kelvin had to do in my book *A Heat Transfer Textbook* (Englewood Cliffs, N.J.: Prentice-Hall, Inc., 1987), Section 5.5.

8   H. Adams, *The Education of Henry Adams* (New York: The Heritage Press, 1918, or one of the many subsequent printings). See especially Chapter 25 and, to some extent, the subsequent chapters.

CHAPTER 3

1   For a fine set of examples of unexpected outcomes, see J. E. Cohen, "The Counter-intuitive in Conflict and Cooperation," *American Scientist*, 76, 6 (1988): 576-84. I talk about the thickness-of-insulation result in my book *A Heat Transfer Textbook* (Englewood Cliffs, N.J.: Prentice-Hall, Inc., 1987), Section 5.5.

2   For the story of Christopher Wren and the dome at St. Paul's, see N. Stirling, *Wonders of Engineering* (Garden City, N.Y.: Doubleday and Company, Inc., 1966).

3   Far and away the most complete history of Count Rumford is Sanford C. Brown's biography, *Benjamin Thompson, Count Rumford* (Cambridge, Mass.: The MIT Press, 1981).

4   Lines from the unpublished work "Antiphon for Evariste Galois" are used with the permission of Carol Christopher Drake.

5 The history of both Evariste Galois and Georg Cantor may be found in C. C. Gillespie, ed., *The Dictionary of Scientific Biography* (New York: Charles Scribner's Sons, 1974).

6 A remarkably rich account of Gibbs' seemingly gray life was written by the noted American poet M. Rukeyser. See *Willard Gibbs* (Garden City, N.J.: Doubleday Duran and Co., Inc., 1942).

7 A simple account of Snow's work is to be found in R. Dubos, and M. Pines, *Health and Disease* (New York: Time-Life Books, 1965). For a fascinating fictionalized account of John Snow and the Broad Street well, see L. P. Taylor, *The Drummer Was the First to Die* (New York: St. Martin's Press, 1992).

8 M. Cheney, *Tesla: Man Out of Time*. (New York: Bantam Doubleday Dell Publishing Group, Inc., 1981).

9 The stories of Jakob and Boelter are told in E. Jakob, "Max Jakob, July 20, 1879–January 4, 1955: Fifty Years of his Work and Life," and F. Kreith, "Dean L. M. K. Boelter's Contribution to Heat Transfer as Seen Through the Eyes of His Former Students." Both articles may be found in E. T. Layton and J. H. Lienhard, eds., *History of Heat Transfer: Essays in Honor of the 50th Anniversary of the ASME Heat Transfer Division* (New York: ASME, 1988), 87–116 and 117–37.

10 F. Howard, *Wilbur and Orville: A Biography of the Wright Brothers*. (New York: Ballantine Books, 1987). The Wilber Wright quotation is on pg. 38.

11 J. T. MacGregor-Morris, *The Inventor of the Valve* (London: The Television Society, 1954).

12 *The Dictionary of National Biography* includes a short, but quite detailed and complete biography of Thomas Octave Murdoch Sopwith. *The Dictionary of National Biography* XVIII (Oxford: Oxford University Press, 1949-50), 671–672.

13 The story of continuous-aim firing is told in at least two places: E. E. Morison, *Men, Machines, and Modern Times*. (Cambridge, Mass: MIT Press, 1995), Chapter 2, "Gunfire at Sea: A Case Study of Innovation," and R. G. Robertson "Failure of the Heavy Gun at Sea: 1898–1922," *Technology and Culture*, 28, 3 (1987): 539-57.

CHAPTER 4

1 The Liquid Paper story is told in E. A. Vare and G. Ptacek, *Mothers of Invention* (New York: Quill, 1987).

2 I take my material on children and bicycles from J. Piaget, *The Child's Conception of Physical Causality* (Totowa, N.J.: Littlefield, Adams & Co., 1969), Sect. III, "Explanation of Machines."

3 J. Porteous, *Coins*. (London: Octopus Books Ltd., 1964).

4 For a good account of coal, see J. Gimpel, *The Medieval Machine: The Industrial Revolution of the Middle Ages* (New York: Penguin Books, 1976), especially Chapter 4.

5 P. C. Welsh, *Woodworking Tools: 1600-1900* (Washington, D.C.: Smithsonian Institution, 1966).

6 S. Lubar, "Culture and Technological Design in the 19th-Century Pin Industry: John Howe and the Howe Manufacturing Company," *Technology and Culture*, 28, 2 (1987): 253-82.

7 A strong impression of the Republic of Texas is created by E. N. Murry, *Notes on the Republic* (Washington, Tex.: Star of the Republic Museum, 1991).

8 W. Reyburn, *Flushed with Pride: The Story of Thomas Crapper* (Englewood Cliffs, N.J.: Prentice-Hall, Inc., 1971).

CHAPTER 5

1 The emergence of baroque architecture is discussed by S. L. Sanabria, "From Gothic to Renaissance Stereotomy: The Design Methods of Philip de l'Orme and Alonso de Vandelvira," *Technology and Culture* 30, 2 (1989): 266–299.

2 R. Burton, *The Anatomy of Melancholy*, abridged and edited by Joan K. Peters (New York : F. Ungar Publishing Co., 1979).

3 Galileo's tower experiment is described in many translations. I used *Two New Sciences*, translated and annotated by Stillman Drake (Madison: University of Wisconsin Press, 1974), "Second New Science, of Local Motions."

4 The analysis of Galileo's tower experiment was done in T. B. Settle, "Galileo and Early Experimentation," in F. Aris, H. T. Davis, and R. H. Stuewer, eds., *Springs of Scientific Creativity*, (Minneapolis: University of Minnesota Press, 1983), 3–20. For more on Galileo, check the following fine Web site: http://es.rice.edu:80/ES/humsoc/Galileo/.

5 For more on Guericke, see A. P. Usher, *A History of Mechanical Inventions*. (Cambridge, Mass.: Harvard University Press, 1970), Chapter 13. It is useful to read an account of Guericke's experiments from only a century and half after they were made. See D. Lardner, *Hydrostatics and Pneumatics*, American ed. (Philadelphia: Carey and Lea, 1832), Part II, "Pneumatics," Chapter 3.

6 For a good overview history of early clockwork, see Usher, *A History of Mechanical Inventions*, Chapter 12.

7 For more on Galileo and the pendulum clock as well as for material on Robert Hooke, see A. Wolf, *A History of Science, Technology, and Philosophy in the 16th and 17th Centuries* (London: George Allen & Unwin Ltd., 1950).

8 M. Espinasse, *Robert Hooke* (Berkeley: University of California Press, 1962). We have no picture of Hooke, but G. Keynes, in *A Bibliography of Dr. Robert Hooke* (Oxford: Clarendon Press, 1960), notes that might be because he was ugly and didn't want his image recorded. His biographer, Aubrey, says that he was "but of midling stature, something crooked, pale faced, and his face but little below, but his head is lardge; his eie full and popping, and not quick; a gray eie. He has a delicate head of haire browning and of an excellent moist curle." Another contemporary, Richard Walker, said of him, "As to his person he was most despicable, being very crooked."

9 L. T. C. Rolt, *The Aeronauts: A History of Ballooning, 1783–1903* (New York: Walker and Company, 1966).

10 J. Tyndall, *Sound*, 3rd ed. (New York: Greenwood Press, 1969). This is a reprint of the 1903 edition.

11 For a clear and simple account of Hale and his telescopes see, N. Stirling, *Wonders of Engineering* (Garden City, N.J.: Doubleday and Company, Inc., 1966), Chapter 11.

12 The Second Law of Thermodynamics is detailed in any undergraduate thermodynamics text. I especially recommend W. C. Reynolds and H. C. Perkins, *Engineering Thermodynamics* (New York: McGraw-Hill Book Co., 1977), Chapter 6.

13 My favorite sources for the principle of Le Chatelier and Braun are P. S. Epstein, *Textbook of Thermodynamics* (New York: John Wiley & Sons, Inc., 1937) Chapter 21;

and H. B. Callen, *Thermodynamics and an Introduction to Thermostatistics*, 2nd ed. (New York: John Wiley & Sons, 1985), Sections 8.4 and 8.5.

CHAPTER 6

1 The full title of Diderot's encyclopedia was *Encyclopédie ou dictionnaire raisonné des sciences, des arts et des métiers*. Check your library for various original or facsimile editions.

2 For the story of Parcieux's work on the Crécy waterwheel, see T. S. Reynolds, "Scientific Influences on Technology: The Case of the Overshot Waterwheel, 1752-1754," *Technology and Culture* 20, 2, (1979): 270-95.

3 For more on the Lunar Society, see J. Bronowski, *The Ascent of Man* (Boston: Little, Brown and Company, 1973), Chapter 8, "The Drive for Power"; and R. E. Schofield, *The Lunar Society at Birmingham: A Social History of Provincial Science and Industry in Eighteenth-century England*, (Oxford: Clarendon Press, 1963).

4 J. Nicholson, *The Operative Mechanic and British Machinist; Being a Practical Display of the Manufactories and Mechanical Arts of the United Kingdom.* (London: 1825). Second American edition, Philadelphia: James Kay, Jun. & Co. Printers, 1831.

5 J. Kanefsky and J. Robey, "Steam Engines in 18th-Century Britain: A Quantative Assessment," *Technology and Culture*, 21, 2 (1980): 161-86.

6 F. T. Evans, "Roads, Railways, and Canals: Technical Choices in 19th-Century Britain," *Technology and Culture*, 22, 1 (1981): 1-34.

CHAPTER 7

1 The *Encyclopaedia Britannica* offers a good starting point on Vespucci. See also J. N. Wilford, *The Mapmakers* (New York: Vintage Books, 1982); D. J. Boorstin, *The Discoverers: A History of Man's Search to Know His World and Himself* (New York: Random House, 1983), Chapter 33.

2 The story of the "discovery" of oxygen is told, along with its implications in T. Kuhn, *The Structure of Scientific Revolutions*, 2nd ed. (Chicago: University of Chicago Press, 1970).

3 For more on canoes, you are again well advised to begin with the *Encyclopaedia Britannica* article on them. For more on the Aleut kayak, see the excellent book by G. Dyson, *Baidarka: The Kayak* (Anchorage: Alaska Northwest Books, 1986).

4 E. N. Hartley, *Ironworks on the Saugus: The Lynn and Graintree Ventures of Undertakers of the Ironworks in New England* (Norman: University of Oklahoma Press, 1957). To see the rebuilt Saugus (or Hammersmith) Iron Works, you may look at the National Park Service's Web site on the Saugus works: http://www.nps.gov/sair/.

5 P. Rouse Jr., *The Printer in Eighteenth-Century Williamsburg* (Williamsburg: Colonial Williamsburg, 1974), revised from the original 1955 printing.

6 For the story of the first American steam engine, see C. W. Pursell Jr., *Early Stationary Steam Engines in America: A Study in the Migration of a Technology* (Washington, D.C.: Smithsonian Institution Press, 1969), Chapters 1 and 4.

7 I. B. Cohen, *Benjamin Franklin's Science* (Cambridge, Mass: Harvard University Press, 1990), Chapter 9.

8 H. L. Abbot, *Beginning of Modern Submarine Warfare*, ed. Frank Anderson (Hamden, Conn.: Archon Books, 1966; facsimile of an 1881 pamphlet); A. Roland,

"Bushnell's Submarine: American Original or European Import?" *Technology and Culture*, 18, 2 (1977): 157-74.

9 For more on the mood of America just before and during the Revolution see, for example, K. Silverman, *A Cultural History of the American Revolution: Painting, Music, Literature and the Theatre in the Colonies and the United States from the Treaty of Paris to the Inauguration of George Washington, 1763-1789* (New York: Columbia University Press, 1987).

10 K. Clark, *Civilisation* (New York: Harper & Row, 1969), p. 264.

11 For more on Oliver Evans, see not only Pursell, *Early Stationary Steam Engines,* but also J. T. Flexner, *Steamboats Come True,* 2nd ed. (Boston: Little, Brown and Company, 1978). Flexner also discusses Fulton's catamaran.

12 A good account of the Erie Canal is given in J. Tarkov, "Engineering the Erie Canal," *American Heritage of Invention & Technology,* summer 1996, 54-57.

13 Anonymous, *Pittsburgh in 1816* (Philadelphia: Carnegie Library, 1916).

14 J.T. Flexner, *Steamboats Come True* (Boston: Little, Brown and Company, 1944, 1978) Chapter 26.

15 S. Lubar, *Engines of Change* (Washington, D.C.: Smithsonian Institution, 1986).

16 T. Dublin, *Lowell: The Story of an Industrial City* (Washington, D.C.: Division of Publications, National Park Service, 1992).

CHAPTER 8

1 Chinese experiments with humans flying in kites are discussed in R. Temple, *The Genius of China* (New York: Simon & Schuster, 1986), Chapter 9.

2 The Egyptian "model airplane" is described in I. van Sertima, ed., *Blacks in Science: Ancient and Modern* (New Brunswick, N. J.: Transaction Books, 1983, 1998), 92-99.

3 The modern flight of the human-powered *Daedelus* has been told in several contemporary sources. See, for example, C. Gorman, "On the Wings of Mythology," *Time,* May 2, 1988, 67; J. Kluger, "Human-Powered Flight," *Discover,* January 1989, 70-71; J. S. Langford, "Triumph of Daedalus," *National Geographic,* August 1988, 190-99; I. Peterson, "On a Wing and a Pedal," *Science News,* April 30, 1988, 277.

4 For good accounts of the early balloonists, see T. D. Crouch, *The Eagle Aloft: Two Centuries of the Balloon in America* (Washington, D.C.: Smithsonian Institution Press, 1983); D. D. Jackson, *The Aeronauts* (Alexandria, Va: Time-Life Books, 1981); L. T. C. Rolt, *The Aeronauts: A History of Ballooning, 1783-1903* (New York: Walker and Company, 1966).

5 Rolt, *The Aeronauts,* p. 72.

6 A fine account of the history of early civil aircraft including helicopters and flying boats, is given in E. Angelucci, *World Encyclopedia of Civil Aircraft* (New York: Crown Publishers, Inc., 1982), Chapter 1.

CHAPTER 9

1 The story of three-field crop rotation and its role in European history is told in L. White Jr., *Medieval Technology and Social Change* (New York: Oxford University Press, 1966), especially Chapter 2.

2 For the history of mechanical clocks, see A. P. Usher, *A History of Mechanical Inventions* (London: Oxford University Press, 1970), Chapter 7.

3 The story of Ceridi's Archimedean pump is told in B. S. Hall and D. C. West, eds.,

Humana Civilitas: Sources and Studies Relating to the Middle Ages and the Renaissance, vol. I, *On Pre-Modern Technology and Science* (Malibu: Undena Publications, 1976).

4 For more on the matter of tunneling, see Jeffrey K. Stine and Howard Rosen, eds., *Going Underground* (Kansas City, Mo: American Public Works Association, 1998). This is the proceedings of a workshop for the Smithsonian Institution's symposium on its tunneling exhibition, "Down Under: Tunnels Past, Present, and Future," National Museum of American History, Saturday, October 23, 1993.

5 D. Lardner, *Popular Lectures on the Steam Engine, in which its Construction and Operation are familiarly Explained; with an Historical Sketch of its Invention and Progressive Improvement* (New York: Elam Bliss, 1828).

6 F. L. Holmes, "Justus von Liebig," *Dictionary of Scientific Biography*, vol. 8 (New York: Charles Scribner's Sons, 1973), 329-50.

7 E. A. Ronnberg Jr., "A Few Words About This Picture," *American Heritage of Invention and Technology*, Fall 1988, pp. 14-20.

8 The classic book on the sinking of the *Titanic* was W. Lord, *A Night to Remember* (New York: Holt, Rienhart and Winston, 1976). I also found that this thrown-together, edited volume, published thirty-three days after the sinking of the *Titanic*, was particularly illuminating both for what it says and what it fails to say: M. Everett, ed., *Wreck and Sinking of the Titanic*. (L. H. Walter, 1912).

CHAPTER 10

1 For an excellent discussion of just what life's end means and how it comes about, see S. Nuland, *How We Die: Reflections on Life's Final Chapter* (New York: Alfred A. Knopf, 1994).

2 The Japanese balloon bomb and its use in World War II is discussed in J. McPhee, "Balloons of War," *The New Yorker*, January 29, 1996, 52-60.

3 I provide extensive data in support of the imperviousness of creative improvement of technologies to outside influence in J. H. Lienhard, "Some Ideas about Growth and Quality in Technology," *Technological Forecasting and Social Change*, 27 (1985): 265-281.

4 The converse to the argument that war drives technology is built by Martin van Creveld. He weaves the more credible argument that the form and shape of war is strongly formed by the availability of technology. See M. van Creveld, *Technology and War: From 2000 BC to the Present* (New York: The Free Press, 1989.)

5 B. S. Hall and D. C. West, eds., *Humana Civilitas: Sources and Studies Relating to the Middle Ages and the Renaissance*, vol. 1, *On Pre-Modern Technology and Science* (Malibu: Undena Publications, 1976).

6 A. H. G. Fokker and B. Gould, *Flying Dutchman: The Life of Anthony Fokker* (New York: Arno Press, 1972).

7 B. W. Tuchman, *The Guns of August* (New York: Macmillan, 1962).

8 W. B. Robinson, *American Forts: Architectural Form and Function* (Urbana: University of Illinois Press, 1977).

9 S. A. Earle, "Life Springs from Death in Truk Lagoon," *The National Geographic Magazine*, May 1976, 578-611.

10 J. R. Chiles, "The Ships That Broke Hitler's Blockade," *American Heritage of Invention & Technology* 3, 3 (1988): 26-32, 41.

11 R. Wohlstetter, *Pearl Harbor: Warning and Decision* (Stanford, Calif.: Stanford

University Press, 1962); R. H. Worth, Jr., *Pearl Harbor* (Jefferson, N.C.: McFarland & Company, Inc., 1943), Section 3, "Radar: The Great Missed Opportunity."

12 A. Kershaw, *A History of the Guillotine* (New York: Barnes & Noble, Inc., 1993).

13 M. Cheney, *Tesla: Man Out of Time* (New York: Dell, 1981), especially Chapter 5; J.J. O'Neill, *Prodigal Genius: The Life of Nikola Tesla* (New York: David McKay Co., 1944).

CHAPTER 11

1 For more on wheels, see V. G. Childe, "Rotary Motion," in S. Singer, E. J. Holmyard, and A. R. Hall, eds., *A History of Technology*, vol 1, *From Early Times to the Fall of Ancient Empires* (New York: Oxford University Press, 1954), Chapter 9. See also T. I. Williams, *The History of Invention* (New York: Facts on File Publications, 1987), Chapter 4.

2 White discusses the history of the crank in L. White, Jr., *Medieval Technology and Social Change* (New York: Oxford University Press, 1966), Chapter 3.

3 For the windmill, see B. George, "Reaping the Wind," *American Heritage of Invention & Technology* 8, 3 (1993): 8-14. E. Kealey, *Harvesting the Air: Windmill Pioneers in Twelfth-Century England* (Berkeley: University of California Press, 1987), Chapter 7. O. Mayr treats the role of feedback in windmills in *The Origins of Feedback Control* (Cambridge, Mass.: MIT Press, 1970).

4 A nice account of the DC-3 is provided in F. Allen, "The Letter That Changed the Way We Fly," *American Heritage of Invention & Technology*, fall 1998, 6-13.

5 B. Newhall, *The History of Photography* (New York: The Museum of Modern Art, 1964).

6 Ulrich Grigull provides two excellent treatments of the evolution of thermometry. The first provides background on Fahrenheit: "Fahrenheit, a Pioneer of Exact Thermometry," in TK, *Heat Transfer, 1966: The Proceedings of the 8th International Heat Transfer Conference* (Washington, D.C.: Hemisphere Publishing Corp., 1986), 1:9-18. Grigull also discusses the technical evolution of thermometer scales: "Newton's Temperature Scale and the Law of Cooling," *Wärme und Stoffübertragung*, 18 (1984): 195-99.

7 M. Gorman, "Sir William O'Shaughnessy, Lord Dalhousie, and the Establishment of the Telegraph System in India," *Technology and Culture* 12, 4 (1971): 581-601.

8 A short account of the jacquard loom and the connection to Babbage is given in D. S. L. Cardwell, *Turning Points in Western Technology* (Canton, Mass.: Watson Publishing International, 1972), 119-121.

9 For the role of Hollerith cards in the evolution of modern computers, see K. S. Reid-Green, "The History of Census Tabulation," *Scientific American*, February 1989, 98-103.

10 L. Owens, "Vannevar Bush and the Differential Analyzer: The Text and Context of an Early Computer," *Technology and Culture* 27, 1 (1986): 63-95.

CHAPTER 12

1 The French invention of manufacturing with interchangeable parts is told in K. Alder, "Innovation and Amnesia: Engineering Rationality and the Fate of Interchangeable Parts Manufacturing in France," *Technology and Culture*, 38, 2 (1997): 273-311.

2 For manufacturing with interchangeable parts in America, see: D. A. Hounshell, *From the American System to Mass Production, 1800-1822: The Development of Manufacturing Technology in the United States* (Baltimore: Johns Hopkins University Press, 1984).

3 The story of Cadillacs in England is told in M. D. Hendry, *Cadillac, Standard of the World: The Complete Seventy-Year History* (New York: Automobile Quarterly Publications/Dutton, 1973).

4 D. A. Hounshell, "Ford Eagle Boats and Mass Production During World War I," in M. R. Smith, ed., *Military Enterprise and Technological Change* (Cambridge, Mass.: MIT Press, 1987).

5 Wise, G., "Heat Transfer Research in General Electric. 1910-1960: Examples of the Product Driven Innovation Cycle," *History of Heat Transfer: Essays in Honor of the 50th Anniversary of the ASME Heat Transfer Division.* E. T. Layton and J. H. Lienhard, eds. (New York: The American Society of Mechanical Engineers, 1988), pp. 189-211.

CHAPTER 13

1 K. Clark, *Civilisation: A Personal View* (New York: Harper & Row, 1969), Chapter 13, "Heroic Materialism."

2 Le Corbusier, *Aircraft* (New York: Universe Books, 1988).

3 F. H. Steiner, "Building with Iron: A Napoleonic Controversy," *Technology and Culture*, 22, 4 (1981): 700-724.

4 J. R. Chiles, "A Cable Under the Sea," *American Heritage of Invention & Technology*, fall 1987, 34-41.

5 V. Barr, "Alexandre Gustave Eiffel: A Towering Genius," *Mechanical Engineering*, February 1992, 58-65; J. A. Keim, *La Tour Eiffel* (Paris: Editions Tel, 1950); M. P. Levy, "Structure and Sculpture," in J. H. Schaub and S. K. Dickison, eds., *Engineering and Humanities* (Malabar, Fla.: R. E. Krieger Pub. Co., 1987), Section 3.3.

6 T. F. Peters, "The Rise of the Skyscraper from the Ashes of Chicago," *American Heritage of Invention & Technology*, fall 1987, 14-23.

7 R. M. Vogel, *Roebling's Delaware & Hudson Canal Aqueducts* (Washington, D.C.: Smithsonian Institution Press, 1971); R. M. Vogel, *Building Brooklyn Bridge: The Design and Construction*, 1867-1883 (Washington, D.C.: Smithsonian Institution Press, 1983); R. M. Vogel, "Designing Brooklyn Bridge," *Annals of the New York Academy of Sciences*, 421 (1983):3-39.

8 E. A. McKay, "Tunneling to New York," *American Heritage of Invention & Technology*, fall 1988, 22-31.

9 N. Sterling, "The Hoover Dam," *Wonders of Engineering* (Garden City, N.Y.: Doubleday and Company, Inc., 1966), 113-125.

10 The NASA crawler-transporters were named a National Historic Mechanical Engineering Landmark by the American Society of Mechanical Engineers (ASME) on February 3, 1977, at the John F. Kennedy Space Center. The ASME's dedicatory brochure for that ceremony contains additional information.

CHAPTER 14

1 L. Coe, *The Telephone and Its Several Inventors* (Jefferson, N.C.: McFarland & Company, Inc., 1995).

2 Anonymous, *The Book of Ecclesiasticus*, chapter 44, verse 1.

3 J. T. Flexner, *Steamboats Come True*, 2nd ed. (Boston: Little, Brown and Company, 1978).

4 G. R. M. Garratt, "Telegraphy," in C. Singer, E. J. Holmyard, A. R. Hall, and T. I. Williams, eds., *A History of Technology*, vol. 4, *C. 1750–1850* (New York: Oxford University Press, 1958), 644–71.

5 The story of the electric light is told in many places. A very good summary is given in the *Encyclopaedia Britannica* under "Lighting."

6 J. J. Flink, "Innovation in Automotive Technology," *American Scientist*, 73 (1985): 151–61.

7 L. Bryant, "The Development of the Diesel Engine," *Technology and Culture*, 17, 3 (1976): 432–46. A. P. Chalkley, *Diesel Engines for Land and Marine Work*, with an introductory chapter by the late Dr. Rudolf Diesel. (New York: D. Van Nostrand Company, 1917).

8 H. R. Collins, *Red Cross Ambulance of 1898 in the Museum of History and Technology* (Washington, D.C.: Smithsonian Institution, 1965); T. W. van Beck, "The 1,100-Year History of the Ambulance," *The American Funeral Director*, May 1992, 44-63.

9 J. Y. Lee, "Anatomy of a Fascinating Failure," *American Heritage of Invention and Technology*, summer 1987, 55-60.

10 T. D. Crouch, "The Feud Between the Wright Brothers and the Smithsonian," *American Heritage of Invention & Technology*, spring 1987, 34-46;

11 E. Angelucci, *World Encyclopedia of Civil Aircraft: From Leonardo da Vinci to the Present* (New York: Crown Publishers, 1982).

12 B. Markham, *West with the Night* (San Francisco: North Point Press, 1983).

13 T. S. Kuhn, *The Structure of Scientific Revolutions* (Chicago: University of Chicago Press, 1970).

CHAPTER 15

1 J. V. Matson, *The Art of Innovation: Using Intelligent Fast Failure* (University Park, Penn.: Pennsylvania State University, 1991); J. V. Matson, *How to Fail Successfully* (Houston, Tex: Dynamo Publishing Co., 1991).

2 H. Petroski, *Engineers of Dreams: Great Bridge Builders and the Spanning of America* (New York: Alfred A. Knopf, 1995), Chapter 3.

3 *The Hammurabi Code, and the Sinaitic Legislation, with a Complete Translation of the Great Babylonian Inscription Discovered at Susa, by Chilperic Edwards* (Port Washington, N.Y.: Kennikat Press, 1971).

4 I tell the tall tale about freezing in J. Lienhard, *A Heat Transfer Textbook*, 2nd ed. (Englewood Cliffs, N.J.: Prentice-Hall, Inc., 1987) 153.

5 K. E. Bailes, "Technology and Legitimacy: Soviet Aviation and Stalinism in the 1930s," *Technology and Culture*, 17, 1 (1976): 55-81.

6 I. B. Holley Jr., "A Detroit Dream of Mass-Produced Fighter Aircraft: The XP-75 Fiasco," *Technology and Culture*, 28, 3 (1987):578-93. For more on the XP-75, see B. Yenne, *The World's Worst Aircraft* (New York: Barnes & Noble Books, 1987), 66-67.

7 P. Likens, "Spacecraft Attitude Dynamics and Control—A Personal Perspective on Early Developments," Invited Lecture preprint, American Institute of Aeronautics and Astronautics (1985).

8 The story of the Hyatt skywalk failure is told in H. Petroski, *To Engineer Is Human: The Role of Failure in Successful Design* (New York: St. Martin's Press, 1985).

CHAPTER 16

1 R. M. Pirsig, *Zen and the Art of Motorcycle Maintenance* (New York: Bantam Books, 1975).

2 For more on Thoreau as an engineer, see H. Petroski, *The Pencil: A History of Design and Circumstance* (New York: Knopf, 1990).

3 For the Robert Burns quotation, see K. Clark, *Civilisation: A Personal View* (New York: Harper & Row, 1969), Chapter 13, "Heroic Materialism"; For the Shelley, Scott, and Blake quotations, see any of the many collections of those poets' work, e.g., P. B. Shelley, *The Complete Poetical Works of Percy Bysshe Shelley*, ed. G. E. Woodberry (Cambridge: Riverside Press, 1901); Sir Walter Scott, *The Complete Poetical Works of Sir Walter Scott*, ed. H. E. Scudder (Cambridge: The Riverside Press, 1900); *The Complete and Selected Prose of John Donne & the Complete Poetry of William Blake*, with an Introduction by Robert Silliman Hillyer (New York: Random House, 1941).

4 M. Shelley, *Frankenstein, or the Modern Prometheus*, ed. M. K. Joseph (London: Oxford University Press, 1969).

5 R. A. Heyman, *Charles Babbage: Pioneer of the Computer* (Princeton, N.J.: Princeton University Press, 1982); J. Bernstein, *The Analytical Engines: Computers, Past, Present, and Future* (New York: Random House, 1964).

6 N. Shute, *No Highway* (New York: William Morrow, 1948). The other Shute books mentioned in this chapter have been reprinted by different publishers and even under various titles. I recommend you check your library's on-line catalog for his works. You will be in for some fun. Henry Petroski also talks about *No Highway* and the Comet failures in H. Petroski, *To Engineer Is Human: The Role of Failure in Successful Design* (New York: St. Martin's Press, 1985), Chapter 14.

CHAPTER 17

1 C. S. Fischer, and G. R. Carroll, "Telephone and Automobile Diffusion in the United States, 1902-1937," *American Journal of Sociology*, 93, 5, (1988): 1153-78.

2 C. Monaco, "The Difficult Birth of the Typewriter," *American Heritage of Invention and Technology*, spring/summer 1988, 10-21.

3 G. R. Cooper, *The Sewing Machine: Its Invention and Development*, 2nd ed. (Washington, D.C.: Smithsonian Institution Press, 1976).

4 The thermodynamics book in my writing case was P. Epstein, *A Textbook of Thermodynamics*

Index

Diderot, Denis, 88
diesel engines, 201–202, *202*
diet, effect of technology on, 5
diodes, development of, 50
dirigibles, 120–122, *122,* 206
diseases, 31, 44
Don Quixote (Cervantes), 156
Dondi, Jacopo di, 129
doors, development of, 155
Drake, Carol Christopher, 40
drills, development of, 155
Dunant, Jean, 202–203
dynamic similitude, 8

Earle, Sylvia, 147–148
East India Co., 100
East River, bridge over, 188–189
École Polytechnic, 180, 187
Eddystone Lighthouse, 15–16
Edison, Thomas, 45, 50–51, 134, 152, 199–200
education, 4, 181–182, 212, 238
Eiffel, Alexandre Gustave, 185–186
Eiffel Tower, 181, 185–186, *186*
Eilmer of Wiltshire Abbey, 22
Eindecker airplane, 145
Einstein, Albert, 21, 43, 46, 237–238
electricity, 44–45, 104, 151–152, 187
elevators, necessity of, in skyscrapers, 187
Eliot, George (Mary Ann Evans), 69
Empire State Building, 187
encyclopedias, 88–89
engineering, 10–11, 166
Engines of Change (Lubar), 112
England compared with France, 119
English Channel, 118, 131
ENIAC, 52–53, 166
Erie Canal, song about, 109, 110
ethical issues of technology, 140–152
Euler, Leonhard, 89
Evans, Oliver, 94–95, 108, 112–113
Explorer I satellite, *216,* 216–217
explosives, 108

Fabre, Henri, 124
factories, monasteries as, 24
Fahrenheit temperature scale, 6, 161–162
failure as integral part of success, 198, 209
Falklands War, 52
farming, 5, 6, 127–128, 157
Farnham Company, 204
Faust (Goethe), 151, 223–224
Fermi, Enrico, reliance on Gibbs' work, 43
feudal system, Black Death's effect on, 31
Field, Cyrus, 184

film, (photographic), 159–161
firearms, 38–39, 53–54, 144–145, 146, 173, *173*
Firnas, Ibn, 23
First National Bank Building, 186–187
Fitch, John, 105, *106,* 197–198
Flatiron Building, *188*
Fleming, John Ambrose, 50–51
flight. *See also* airplanes; balloons; dirigibles;
    pilots
    ballooning records, 214
    childhood dreams of, 115
    development of, 49–50
    early glider flights, 22–23
    heavier-than-air, 122–123
    transatlantic flights, 206–207
Flink, James, 200
fluid flow, study of, 18
Flushed with Pride (Reyburn), 67–68
Fokker, Anthony, 144–145
Ford, Henry, 174, 176
Forlanini, Enrico, 123
Fort Sumter, 146
fossil fuels, 62, 133
Fourier, Joseph, 32
France compared with England, 119
Frankenstein (Shelley), 222–224
Franklin, Benjamin, 90, 97, 102–103, 118
French Revolution, 86, 88–89, 196–197
Fulton, Robert, 97, 108–112, 195, 198

Galileo Galilei, 70–74, 161
Gallatin, Albert, 109
Galois, Evariste, 39–40
Gay-Lussac, Joseph Louis, 134
General Electric Co., 177
General Motors, airplane production, 215
genetics in development of wheat, 5–6
Gentleman's Magazine, 104–105, *105*
geology, calculating age of planet, 32–33
Gibbs, James, 233
Gibbs, Josiah Willard, 42–43
Giffard, Henri, 120
gimlet auger, 62
glass, 59–61, 103, 110
gliders, 49
Godwin, Mary, 222
Goethe, Johann Wolfgang von, 151
googol (number), 41
Gossamer Albatross, 132–133
Graham, Sylvester, 66
Grand Coulee, 90
graphics, 17–18
grass, *6*
Great Britain manufacturing, 174–175